I0054630

Environment and Health in Jammu & Kashmir and Ladakh

This engaging book presents an insightful look into the contributing factors that have shaped the modern public health system in Jammu & Kashmir and Ladakh. Reflecting on the historical, socio-economic, and contemporary scenario of environment and public health, this book presents chapters that discuss the role of spatial patterns of diseases, health-risk patterns, contributions of medical missionaries in health services in Kashmir, changing disease ecology of Leh, and traditional medical therapy in Ladakh, among others.

This book also examines the cholera ecology in Kashmir during the 19th century, and the significance of Kangri – a portable traditional heat source – in cultural studies, economics, and cancer research. It investigates the role of traditional knowledge in the medical therapy of rural areas of Ladakh and the impact of urbanization on the quality of human health in Srinagar City. Besides, this book examines iodine deficiency disorders and the extension of vector-borne diseases. The essays also probe into the rising mental health concerns in post-pandemic Kashmir.

This book will be useful for students, researchers, and teachers of physical geography, human geography, environmental studies, public health, and health sciences. It will also be of interest to political geographers, sociologists, policymakers, and those interested in the issues related to health and environment in the region.

Prof. Rais Akhtar has taught at the Jawaharlal Nehru University, New Delhi; the University of Zambia, Lusaka; and the University of Kashmir, Srinagar, Jammu & Kashmir. He is the first geographer who published his research in medical geography and climate change and health in national and international journals, including the prestigious medical journal *The Lancet*. He has published to his credit more than 100 research papers and 21 books in India and abroad, including Greenwood Press, New York; Harwood Academic publisher, London; and Springer. His books include *Health and Disease in Tropical Africa: Medical and Geographical Viewpoints* (London, 1987), *Health Care Patterns and Planning in Developing Countries* (1990), *Malaria in South Asia* (2010), *Climate Change and Human Health Scenario in South and Southeast Asia* (2016), *Climate Change and Air Pollution* (2018), *Extreme Weather Events and Human Health* (2020),

Coronavirus (COVID-19) Outbreaks, Environment and Human Behaviour (2021) and *Coronavirus (COVID-19) Outbreaks, Vaccination, Politics and Society: The Continuing Challenge* (2022). Prof. Rais Akhtar is the recipient of a number of international fellowships including Leverhulme Overseas Visiting Fellow (University of Liverpool); Henry Chapman Fellow, University of London; Commonwealth Secretariat Fellowship, Institute of Commonwealth Studies, London; Visiting Fellow, University of Sussex and London School of Hygiene and Tropical Medicine; Royal Society Short Visiting Fellowship, University of Oxford; Visiting Professor, University of Paris, Nanterre, Paris-10; Visiting Fellowship, Free University of Brussels, Belgium; French Government Visiting Fellowship and MSH Visiting Fellow, Paris; Visiting Fellow, Welcome Trust, London; and Visiting Lecturer, University of Akron, USA.

Environment and Health in Jammu & Kashmir and Ladakh

Historical and Contemporary Scenario

Editor-Rais Akhtar

Routledge
Taylor & Francis Group

LONDON AND NEW YORK

First published 2024
by Routledge
4 Park Square, Milton Park, Abingdon, Oxon OX14 4RN

and by Routledge
605 Third Avenue, New York, NY 10158

Routledge is an imprint of the Taylor & Francis Group, an informa business

© 2024 selection and editorial matter, Rais Akhtar individual chapters, the contributors

The right of Rais Akhtar to be identified as the authors of the editorial material, and of the authors for their individual chapters, has been asserted in accordance with sections 77 and 78 of the Copyright, Designs and Patents Act 1988.

All rights reserved. No part of this book may be reprinted or reproduced or utilised in any form or by any electronic, mechanical, or other means, now known or hereafter invented, including photocopying and recording, or in any information storage or retrieval system, without permission in writing from the publishers.

Disclaimer: The views and opinions expressed in this book are those of the authors and do not necessarily reflect the views and opinions of Routledge. Authors are responsible for all contents in their articles including accuracy of the facts, statements, and citations.

Trademark notice: Product or corporate names may be trademarks or registered trademarks, and are used only for identification and explanation without intent to infringe.

British Library Cataloguing-in-Publication Data
A catalogue record for this book is available from the British Library

ISBN: 978-1-032-34631-1 (hbk)
ISBN: 978-1-032-35947-2 (pbk)
ISBN: 978-1-003-32945-9 (ebk)

DOI: 10.4324/9781003329459

Typeset in Times New Roman
by codeMantra

Contents

Figures

Tables

Contributors

Prof. Rais Akhtar has taught at the Jawaharlal Nehru University, New Delhi; the University of Zambia, Lusaka; and the University of Kashmir, Srinagar, Jammu & Kashmir. He is the first geographer who published his research in medical geography and climate change and health in national and international journals, including the prestigious medical journal *The Lancet.*

Dr. M. Sultan Bhat is presently a Professor in the Department of Geography & Disaster Management, University of Kashmir. He has been actively involved in teaching and research during the last 30 years. He has published 120 research papers and two books to his credit and is guiding research in Urban Geography, Watershed Development, Climate Change, and Environmental Studies. He has guided about 18 MPhil students and 15 PhD scholars for their research degrees. He has availed Baden Wurttemberg Visiting Professor Fellowship at South Asia Institute, Heidelberg University, Germany. He is a NABET-accredited functional area expert (FAE) in socio-economics and has been associated with EIA & EMP and SIA & SMP of a number of hydroelectric projects, roadways, industrial complexes, and township developments of Jammu & Kashmir state. He has also completed a number of research projects sponsored by ICSSR, UGC, DST, and MoES (Ministry of Earth Sciences, Government of India). Dr. Bhat is currently working on a research project entitled "Climatic Variability and Disaster Vulnerability Assessment of Upper Indus Basin, Kashmir Himalayas".

Dr. John Bray is an independent historian, currently based in Singapore. He was the President of the International Association for Ladakh Studies from 2007 to 2015. He has published widely on the history of Christian missions and British diplomacy in the 19th-century and early 20th-century Ladakh.

Dr Rajiv Kumar Gupta is a Professor and Head of Community Medicine, Government Medical College, Jammu. He has more than a hundred publications and three book chapters. He got Vasudeva Oration in 2018 and was nominated for Fellow in PSM in 2022 to be conferred FIAPSM in 2023. Also, he is a fellow in IMSA, on the forefront in COVID-19 mitigation in the UT of J&K, and a regular resource person in DD, AIR, and newspapers.

Talat Jabeen is an Epidemiologist in Bulbul Bagh, Burzulla, SSO, Kashmir.

Mushtaq A. Kumar is a young research scholar pursuing PhD in the Department of Geography and Disaster Management, University of Kashmir. He has published five research papers in the field of medical geography and has also participated and presented research papers in national and international conferences.

Bashir A. Lone is pursuing PhD in the Department of Geography at the University of Kashmir and has qualified for UGC NET.

Sheraz Ahmad Lone is a PhD research scholar in the Department of Geography and Disaster Management, University of Kashmir, Srinagar. He is presently working as a teacher.

Arshad Ahmad Lone, Research Scholar Department of Geography and Disaster Management, University of Kashmir, Srinagar, J&K.

Richa Mahajan is a Lecturer in the Department of Community Medicine, Government Medical College, Jammu, J&K.

Dr. Davinder Singh Manhas is a professor in the Department of Geography, University of Jammu (on a contractual basis) and specializes in resource potential studies and geospatial techniques. He has published a number of research papers.

Prof. Ishtiaq Ahmad Mayer is an Ex-Professor and Head of the Department of Geography, University of Kashmir, Srinagar, and he has published numerous research works that mostly highlighted environment and health issues among the people of Western Himalaya.

Prof. Anuradha Sharma is the Head of the Department of Geography, University of Jammu. She is the President of the Association of Punjab Geographers and Vice President of Deccan Geographers. She specializes in Resource and Gender Studies in Geography and has research projects awarded by the DST.

Dr. Shashi Sudhan Sharms is the Principal and Dean, Government Medical College, Jammu. She is a renowned microbiologist, with several publications and book chapters to her credit.

Dr. Subhash Chander Sharma, MSc, PhD, is a Retd. Associate Professor in Higher Education Department, Government of Jammu and Kashmir. He served ten years in P.G. Geography Department of the University of Jammu on deputation and also worked as the Head of the Department of Geography of various Government Colleges of Jammu Region.

Dr. Sandeep Singh specializes in forest resource studies. He has authored many publications, and currently, he is a teacher in a Higher Secondary School.

Dr. Sunil Raina is a Professor and Head of Community Medicine at Dr. RP Government Medical College, Kangra at Tanda, Himachal Pradesh. He heads the Multi-Disciplinary Research Unit and is the Director of the Centre for Advancing Tobacco Control in Himachal Pradesh (CATCH). He is a member of the Expert Committee of ICMR – Centre for Non-Communicable Diseases.

Dr. G.M. Rather is a senior faculty member at the Department of Geography and Disaster Management, University of Kashmir. He has done his specialization in medical geography, and he is teaching medical geography for the last 15 years. He has published more than 45 research papers in the field of medical geography in international and national journals.

Manzoor A. Wani is an Assistant Professor in higher education with PhD, MPhil, and qualified for UGC JRF-NET in geography. He has published a number of research papers and book chapters, and he has co-authored a book.

Bupinder Zutshi, MA, MPhil, PhD, is a Superannuated Professor from Jawaharlal Nehru University, New Delhi; a Senior Fellow at the Indian Council of Social Sciences Research (ICSSR); and a former Independent Director of Northeastern Electric Power Corporation Limited (NEEPCO). He has more than 40 years of teaching and research experience at postgraduate and MPhil/PhD research level. He has published nine books, has completed more than a dozen research projects, and has written several research articles (more than 40) in several research journals of repute, published in India and abroad. He has organized several national and international seminars and conferences and participated/presented research papers in several national and international seminars. Major research interests are issues related to human settlements and their changing habitat. Other research interests are issues related to demographic changes and their consequences on population composition and population characteristics. He has written books on Disaster Risk Reduction – Community Resilience.

Foreword

This volume, edited by Prof. Rais Akhtar, an eminent academic and researcher and a pioneer in the field of Medical Geography, presents over a dozen of case studies that illustrate how the quality of health and the incidence of disease in the Jammu, Kashmir, and Ladakh regions have been affected by environmental factors. Also brought out is the historical perspective: how the patterns of diseases and health care systems have changed over the past centuries, consequent to changes in the physical, social, cultural, and economic environments.

These case studies are of significant current relevance: they clearly bring home that a sound and systematic understanding of the spatial patterns of diseases (both communicable and non-communicable) and of the geographical environments in which they occurred (even leading to epidemics) in the past periods provides the national health authorities with a veritable basis for forecasting disease patterns in the coming time and to plan accordingly for the training of health workers and for re-orienting the health care delivery systems that would be required to meet the arising challenges.

The health authorities in the Union Territories of Jammu and Kashmir and Ladakh would particularly benefit by studying this volume and appropriately reviewing their existing health care policies and programmes. For instance, one of the case studies in this volume brings out the serious medical problems that are occurring and also causing different kinds of cancers, because of the existing unscientific practice followed by orchardists to carry out repeated pesticide spraying on apple and other fruit trees in the Kashmir Valley. The labours engaged to carry out the multiple spraying operations are not provided with any protective gear, nor even educated about the precautions they must observe while handling the poisonous pesticides. Unless such hazardous practices are effectively regulated by the concerned governmental authorities, poor daily wagers will continue to suffer serious ailments and precious lives shall continue to be lost every year.

Considering the crucial connectivity between Health and Environment, it is hoped that the planning authorities in the Central and State governments shall timely launch necessary initiatives to meet the emerging challenges, particularly keeping in mind the rapid pace at which the phenomenon of global warming and climate change are already advancing. India has several highly diverse geographical zones – mountainous areas, plains, deserts, coastal areas, and island territories – with sensitive bio-diversities that must be protected. Also, we must not forget how, in the past years, the changing weather patterns have resulted in causing tsunamis, tornados, cloud bursts, floods, and other natural disasters causing enormous human and economic losses.

I compliment Prof. Akhtar for ploughing his vast knowledge and that of the other academics whose researches have been brought out in this volume, to focus not on an esoteric theme but on real issues that concern the health and well-being of over 135 crore human beings in our country.

N.N. Vohra
Formerly Governor of J&K State (2008–18)

15 January 2022
Gurugram

Preface

Medical geography is concerned with the analysis of spatial patterns of disease and health care provision. John M. Hunter of Michigan State University correctly defined the field as "the application of geographical concepts and techniques to health-related problems" (Hunter 1974).

Thus, this branch of geography utilizes the concept and techniques of geography to investigate health-related issues. The analysis of the areal or spatial patterns of disease and health care provisions may be often based on mapping of various indicators of health, disease, and health care provisions. Such process helps understand the environmental framework of the occurrences of diseases and may lead to future forecasting of diseases.

Historical medical geography deals with changing environmental conditions – both physical and socio-economic and cultural patterns and the incidence of diseases. It is always fascinating to compare the pattern of diseases and access to health in the past with contemporary scenario of health and disease patterns to identify the process of change in the pattern of health and disease.

Sudden outbreaks of cholera in Kashmir during 1850–92 have been described by many writers and physicians working in Kashmir during that period as indigenous due to insanitary conditions prevailing in Kashmir. It is the contention of this paper that although insanitary conditions were responsible for the spread of the disease, cholera was brought to Kashmir as a result of population mobility between Kashmir and Punjab and was the result of a pandemic wave. This is confirmed by the data and information available on the cholera outbreaks in other parts of the Indian subcontinent, Europe, and the United States. Chapter 2 also examines the cholera ecology in Kashmir during the 19th century.

In 1867, the British Indian government sent Dr. Henry Cayley (1834–1904) on an official mission to Ladakh, and he returned to the region every summer until 1871. His orders were to monitor political and commercial developments in Central Asia. At the same time, since he was a doctor in the Indian Medical Service, he opened a dispensary in Ladakh as a means of making contact with local people as well as traders from other regions. He, therefore, has the distinction of being one of the first Western-trained physicians to practise in Ladakh, and his reports provide valuable insights into health conditions in the 19th-century Ladakh. This essay places Cayley in his historical context, showing how the themes of medicine and diplomacy complemented each other in early British engagement with Ladakh and Central Asia.

Mark Harrison has said that hospitals occupy a central place within health care systems, not only on account of their curative functions but also as centres of teaching and research. The indigenous system of medicine practised by Hakims in Kashmir is the Unani. The Mission Hospital in Srinagar, Kashmir, became the most important hospital attracting patients, not only within Kashmir but also in the surrounding countries and not only because of the curative facilities provided at the hospital but also because of the humane approach of its physicians, outstanding among them being Arthur Neve. The patients represented every class of society. Patients come from villages scattered throughout Kashmir and the Plains of India, and a few from Tibet, Afghanistan, Yarkand, and Khostan. According to Neve, in 1912, there were 23,642 new outpatients and 1979 inpatients. Physical, socio-cultural, and political conditions hinder access to the Mission Hospital. Neve's younger brother Ernest F. Neve (1861–1946) made significant contributions when an earthquake struck and during cholera outbreaks in Kashmir at the end of the 19th century.

Chapter 5 brings out the significance of Kangri in the life of Kashmiri people. It encompasses the environment, culture, economics, and convenience for the people of Kashmir. Cultural tradition and climate necessities work together in the continuance of a centuries-old practice of carrying *Kangri,* a portable heat source, under the clothes during the winter months, from about the beginning of November up until late March. The chapter describes the perspective of *Kangr*i, and it also throws light on the incidence of *Kangri* burn cancer and the research on the cancer done by Dr. W.J. Elmslie, Dr. E.F. Neve, and Imtiaz Wani.

The chapter discusses about the health problems in Kashmir with a focus on principal diseases in the study area. The chapter describes socio-economic conditions and the occurrence of epidemics in the region in the 19th-century Kashmir.

Chapter 7 focuses on the changing disease ecology in Leh district with focus on historical and contemporary scenario. Respiratory tract infection, dental cases, E.N.T, and mumps were major diseases prevalent in Leh.

Traditional knowledge plays an important role in establishing sustainable relationship between man and nature in a society more dependent on natural environment for various needs. This investigation was undertaken with an objective to analyse the role of traditional knowledge in the medical therapy of rural areas of Ladakh. Utilization and preference of traditional medical system, sources of knowledge for utilization of Amchi medical system, spatial distribution and socio-economic background of traditional medical practitioners, perception of disease, and management of health problems were the themes of research. The analysis shows that Amchi – an offshoot of Tibetan medical system is not only used but also preferred in rural areas of Ladakh with traditional source of knowledge. Perception of disease and management of various health problems is also based on traditional knowledge. Some suggestions have been made for future health care planning.

The geographic and climatic attributes of the Kashmir Valley support for various types of horticulture production, which is a major contributor in the economy of its people. The fruit production area of all varieties spread over around 0.2 million hectares of land, of which 0.11 million hectares (more than 50%) are under apple production, engaging about 40% of all orchard farmers. This study has been targeted mainly on such orchardist community in the Kashmir Valley using a variety of pesticides, insecticides, and other fungicides to protect plants, fruits, and leaves from diseases every year. The main objective of the study was to i) document the serious consequences of the pesticide exposure for the health of orchard farmers in Kashmir, ii) document the famers source of information about pesticides, iii) document the signs and symptoms of illness related to pesticide exposure, and iv) ascertain the precautionary measures taken. The sample study revealed that most of the farmers were using unscientific and traditional methods while spraying which puts them on high risk of pesticide exposure resulting in serious health hazards. Most of them were not aware of the health hazards of the inappropriate handling of pesticides. They did not take necessary personal protective measures while handling pesticides. It was observed that sprayers were eating foodstuffs, drinking water, chewing, or smoking while at spraying work, thereby digesting pesticide particles and putting themselves at high health risks. The farmers reported excessive sweating (38.06%), burning/stinging/itching of eyes (36.49%), fatigue (30.31%), dry/sore throat (27.59%), dizziness (27.18%), and numbness/muscle weakness/muscle cramps (24.14%), all more prevalent among sprayers. Awareness to use personal protective measures while handling pesticides is needed among farmers in Kashmir. Farmers need to be encouraged to reduce, if not eliminate, the use of pesticides, with the introduction of incentives to help them shift from synthetic pesticides to bio-pesticides and organic farming.

The main objective of this paper is to analyse the impact of urbanization on the quality of human health in Srinagar City. The study area has been divided into five zones on the basis of household density using the data of the Census of India 2011. The data regarding the quality of health and related facilities of the study area have been processed in Arc View 3.2a for the preparation of various thematic maps. Both primary and secondary data were used and analysed with the help of the composite index technique. The results of the study show greater variation in the quality of human health in the study area. The study concludes that Zone II has very high Quality of Health with a composite index value 18, while a very low level is found in Zone I with a composite value 38.

The erstwhile princely state of Jammu and Kashmir has currently three administrative divisions, viz., Jammu, Kashmir, and Leh. Jammu division is comprised of ten districts that include Jammu, Udhampur, Samba, Kathua, Reasi, Rajouri, Poonch, Ramban, Doda, and Kishtwar. Except Jammu, Samba, and, to some extent, Kathua districts, most of the terrain of Jammu division is mountainous and includes the Pir Panjal Range that separates it from Kashmir Valley. Chenab River is the principal river of the Jammu division. Jammu city, also known as the city of temples, is the largest city in Jammu division and is the winter capital of the state. Major religion of Jammu division is Hinduism (62%) and Islam (36%). The Pir Panjal Range, the Trikuta Hills, and the low-lying Tawi River basin add diversity to the terrain of Jammu division.

The state of Jammu and Kashmir witnesses a range of elevations bringing diversity to its biogeography. As is known, vector-borne diseases (VBDs) are climate sensitive. The reason being that the pathogens involved in VBDs have to complete part of their development in particular species of an insect vector. The change in climate across is expected to extend the presence of VBDs. Therefore, the Jammu and Kashmir state of India with the currently less common presence of VBDs may start recording a higher presence in the future with this climate change.

Extreme weather events are a global phenomenon. Extreme weather events have impacted both natural and human systems. These systems include human health, involving significant social, economic, and environmental consequences. Information and knowledge about the health impacts of extreme weather changes are growing rapidly and are increasingly being shared around the world. This paper examines the impact of the Srinagar flood of 2014 (Jammu & Kashmir) on water-borne disease risks. The study results are based on the primary sample survey of flood victims in Srinagar. Majority of the areas of Srinagar city were completely inundated for seven to ten days during September 2014 floods. The study results confirm that there was a significant increase in waterborne diseases, especially giardiasis, viral gastroenteritis, amoebiasis, and cryptosporidiosis, among the residents of flood-affected areas. Skin rash was a major disease risk suffered by flood victims. The health care system was also under stress due to a significant increase in patients suffering from waterborne diseases and non-functional health services due to the inundation of government hospitals and other private health services.

Health conditions in urban areas are the reflection of the availability of not only medical facilities but also sanitation, transport network, levels of pollution, housing conditions, overcrowding, and recreational facilities. The paper is aimed at the assessment of residents' perception concerning health hazards, which affects the health of the people in different areas. The levels of perception and its impact on people's behaviour vary over time and space. This variation is no doubt related to the socio-cultural and economic background, and the geography of the particular areas. Three localities in Srinagar urban area have been selected in order to study how the residents perceive the health hazards of their locality, identify the variation in their perception if any, and identify what particular health hazard most or least concerned the residuals. It is hoped that such studies based on local field studies may contribute significantly to urban health planning.

India has been identified as one of the top 12 mega bio-diversity centres of the world due to wide variations in climate, soil, altitude, and latitude. The western Himalayan region of of J&K state has a rich heritage of diverse medicinal plants, and these medicinal plants give access and reasonable medicines to poor people. This paper attempts to understand the utilization pattern of these plants with identified selected background characteristics and their health care behaviour in Udhampur.

Understanding of health-seeking behaviour is vital and has been studied by geographers in the framework of socio-economic and cultural dimensions. Cultural aspects play an important role in the health-seeking behaviour as variables like ethnicity, cultural traits, cultural regions, language, religion, and economic aspects have witnessed different health-seeking behaviour. This study was an attempt to study the health-seeking behaviour among the Hanjis – an ethnic group of population inhabiting Dal Lake. The main focus of the study was on different aspects of health-seeking behaviour like the perception of disease, management of health problems, utilization of medical system, and preference of same in relation to socio-economic setup. Suggestions have been made for the improvement of health in this ethnic group.

A significant approach for analysing disease in the context of human–environment interaction is that of disease ecology. It comprehends the spatial pattern of diseases and its correspondence with geographical factors of Jammu region. The study investigates the spatial distribution pattern and levels of diseases on the basis of morbidity percentages by the Rank-coefficient method. Further, the study proposes disease zonation and intensity zones, which provides the foundation to disease regionalization.

Coronavirus (COVID-19) caused global health and economic disaster, which the world has not experienced in recent times. It caused havoc to all geographical regions including the Himalayas. Attempt has been made in this chapter to study the spatial spread of COVID-19, identify high- and low-intensity regions, and discuss various determinants of the distribution pattern of coronavirus in the region.

This study was carried out to access the levels of malnutrition by altitude based on Body Mass Index (BMI) among school children (14–18 years) in Gujjar Community of Greater Kashmir Himalayas. The weight and height of sample children were measured, and the BMI was calculated from the weight and height of all sample children from all altitudinal zones of Kashmir Himalayas. WHO Classification of 2007 was employed for classifying children in grades of malnutrition on the basis of BMI. The study reveals that average weight, average height, and average BMI were less than ICMR standards with good contrast among male and female children with a declining trend with altitude. The main findings depict that the mean height, weight, and BMI were 123.3 cm, 22.105 kg, and 15.96 kg/m^2 for male and 119.6 cm, 21.270 kg, and 14.53 kg/m^2 for female, respectively. Nearly about 16.42% of male children and 16.96% of female children were having BMI <18.5 kg/m^2, while 10.63% of male children and 11.55% of female children were having severe-to-moderate malnutrition. This study shall be of great help for future health planning in this mountainous region.

COVID-19, an emerging viral disease, has caused devastation globally with millions affected and hundreds of thousands of deaths. It has affected all the nations globally and was declared as pandemic in March 2020. The mortality data of both the provinces, viz., Jammu and Kashmir of UT of J&K, were analysed till 31 January 2022. A total of 4,681 deaths were reported during this period with 51% of the total in Kashmir province. About two-third of total deaths were in males, and 50–69 years age group was the worst affected. Jammu district reported 790 deaths per million population, while Srinagar district reported 735 deaths per million population. Co-morbidities were found in 56% of COVID-19 deaths in Jammu province, while the figure for Kashmir Valley was 79%. One-third of deaths were reported within one to three days since admission.

COVID-19, which first marked its presence in December 2019 and continues to prevail till date with oscillating surges and falls, has impacted humans all over the globe. As such, many spheres of human life and activities are being adversely affected by its profound impact on global mental health. The unpredictability of the situation, the uncertainty of when to control the disease, its spread to different geographic locations, and the seriousness of the risk post-infection, all together have aggravated the stressfulness of the situation. There has been an alarming rise in cases of depression and anxiety in the Kashmir Valley, India. In this context, this paper explores the vulnerability of recovered COVID-19 patients to such mental health concerns. The chapter uses primary sources of data with a sample set of 300 recovered COVID-19 patients in one of the largest cities of Kashmir Valley, Srinagar. The sample set is divided into two groups (hospitalized and home quarantined). For depression and anxiety, a questionnaire based on PHQ-9 and GAD-7 is used for all the COVID-19 patients, **37.02%** were found to have clinically defined depression and **24.31%** were found to have clinically defined anxiety. The risk of severe depression and severe anxiety levels is higher in home-quarantined patients (relative risk, RR = 1.308, 95% CI = 0.77–2.2; and relative risk, RR = 1.17, 95% CI = 0.9–1.5, respectively). Three risk factors like retesting positive for COVID-19, pre-existing health conditions (diabetes, hypertension, heart problems), and taking longer to recover from COVID-19 (>12 days) played a significant role in aggravating the severity of mental health problems. The need for mental health support is of utmost importance. An urgent need for researchers, clinicians, and policymakers is the need of the hour to deal with the burning problem of deteriorating mental health in the Kashmir Valley

Finally, this book focuses on the study between historical and contemporary scenario of the environmental perspective of health and disease, which is an enthralling area of research, and highlights the pattern of environmental conditions, people's adjustment with the environment and people's mobility, and the occurrence of health problems as a result of man's interaction with the environment. During the past 160 years, in Kashmir and Ladakh, the contributing factors that have shaped the modern public health system include missionary doctors and the spread of education. It argues the importance of such regional studies in the understanding of environment-associated ailments.

Reference

Hunter J.M. (1974) The challenge of medical geography, in J. M. Hunter (ed.) *The Geography of Health and Disease*, Studies in Geography series, n.6, Dept. of Geography, University of North Carolina at Chappel Hill, pp 1–31.

Acknowledgements

In the process of writing, editing, and preparing this book, there have been many people who have encouraged, helped, and supported us with their skill, thoughtful evaluation of chapters, and constructive criticism.

First of all, I am indebted to all the contributors of the chapters from different parts of Jammu & Kashmir and Ladakh and other places for providing the scholarly and innovative scientific piece of research to make this book a reality. I am also thankful to the two reviewers assigned by our publisher who carefully, timely, and critically reviewed the manuscripts with useful suggestions.

I am also grateful to Shri N.N. Vohra Sahab, former Governor of Jammu & Kashmir (2008–18), for writing the Foreword, which adds greatly to this book with his thoughtful insights on environment and health.

I am thankful to my family – wife Dr. Nilofar Izhar; daughter Dr. Shirin Rais, Assistant Professor, AMU; and my son-in-law Dr. Wasim Ahmad, Associate Professor IIT, Kanpur, who encouraged and sustained me in developing the structure of the book and editing tasks, and I am deeply grateful for their support and indulgence. I am indebted to the *Indian Journal of History of Science (INSA)*, *Journal of Medical Biography of Royal Society of Medicine*, *Focus of American Geographical Society*, *Punjab Geographer*, *Ashgate Publishing*, and *National Association of Geographers, India,* for the permission to reprint my articles (Chapters 2, 4, 5, 7, 14, and 18, respectively). I am also thankful to Dr. Ravi Kumar Bhat, Dr. G.M. Rather, University of Kashmir, and Javed Akhtar, for their assistance. Finally and most essentially, we are deeply obliged to Routledge and the entire publishing team, without whose patience and immense support this book would not have come to fruition. I specially thank Miss Lubna Irfan, Miss Shloka Chauhan, and Miss Shoma Choudhury for their encouraging support.

NOTE: It should be noted that the title of the book mentions Jammu & Kashmir and Ladakh. It actually refers to the erstwhile State of Jammu & Kashmir and Ladakh, which are at present Union Territories.

New Delhi/Aligarh   **RAIS AKHTAR**

About Prof. Rais Akhtar

Prof. Rais Akhtar is presently Adjunct Faculty, International Institute of Health Management and Research, New Delhi. Formerly Professor of Geography, and Dean, University of Kashmir, Srinagar, National Fellow and Emeritus Scientist (CSIR), CSRD, Jawaharlal Nehru University, New Delhi, and Visiting Professor Dept. of Geology, AMU, Aligarh He has taught at the Jawaharlal Nehru University, New Delhi, University of Zambia, Lusaka and The University of Kashmir, Srinagar. He is recipient of a number of international fellowships including Leverhulme Fellowship (University of Liverpool), Henry Chapman Fellowship (University of London), and Visiting Fellowship, (University of Sussex), Royal Society Fellowship, University of Oxford and, Visiting Professorship, University of Paris-10. Prof. Akhtar was elected Fellow of Royal Geographical Society, London and Royal Academy of Overseas Sciences, Brussels.

Prof. Akhtar delivered invited lectures in about 55 universities geography departments and medical colleges including London School of Hygiene and Tropical Medicine, Liverpool School of Tropical Medicine, Dept. of Geography, University of Edinburgh, and School of Public Health, Johns Hopkins University, Baltimore. He was Lead Author (1999-2007), on the Intergovernmental Panel on Climate Change (IPCC), which is the joint winner of Nobel Peace Prize for 2007. Prof.Akhtar is the recipient of Nobel Memento.

Prof. Akhtar has to his credit 91 research papers and 21 Books published from India, United Kingdom, United States, Germany and The Netherlands. His t books entitled:Climate Change and Human Health Scenario in South and Southeast Asia, was published in 2016 by Springer, and his other books include Climate Change and Air Pollution (with C.Palagiano) published by Springer, and Geographical Aspects of Health and Disease in India (with Andrew Learmonth) published in 2018 by Concept (New Delhi). His latest books include: Extreme Weather Events and Human Health, was published by Springer in early 2020, Coronavirus (COVID-19) Outbreaks, Environment and Human Behaviour, and Coronavirus

(COVID-19) Outbreaks, Vaccination, Politics and Society, 2022, were also published by Springer in 2021 and 2022 respectively.

Prof. Rais Akhtar was Member of Expert Group on Climate Change and Human Health of the Ministry of Health and Family Welfare, Govt. of India. Prof. Akhtar has been appointed as Corresponding member (Fellow) of Italian Geographical Society in October 2018.

1 Environment and Health in Jammu & Kashmir and Ladakh

Historical and Contemporary
Scenario

Rais Akhtar

Introduction

Jammu and Kashmir UT of India is located in the northern part of the Indian subcontinent in the vicinity of the Karakoram and westernmost Himalayan mountain ranges (Kirk and Akhtar, 1990). Jammu and Kashmir, formerly one of the largest princely states of India, Jammu and Kashmir is located in the extreme north of the country. The Vale of Kashmir, predominantly populated by Muslims, has also ancient caves and temples of Kashmir that reveal a strong link with Indian culture.

Geographically, this state is divided into four zones – the mountainous and semi-mountainous plain known as Kandi belt, hills including Siwalik ranges, mountains of Kashmir Valley and Pir Panjal Range, and Tibetan tract of Ladakh and Kargil. This state has a number of lakes, rivers, rivulets, and glacial regions. The important rivers of this state are Indus, Chenab, and Sutlej (Jhelum). There are extreme variations in climate in the state, due to its location and topography. The temperature of this state varies spatially. Leh is the coldest and Jammu is the hottest. In winter night in Leh and Jammu, temperatures go down below zero and very often experience snowfall.

The state's total area is 222,236 km², and the area of the Kashmir Valley is 15,520.3 km². The population in 2011 is 12,541,302, and the projected population is estimated to be about 14.324 million in 2018. The population of Srinagar and Jammu cities is 502,197 according to the 2011 census.

Jammu and Kashmir consists of three geographically diverse regions: Jammu, the Kashmir Valley, and Ladakh. Srinagar is the summer capital, and Jammu is the winter capital. Jammu and Kashmir is the only state in India with a Muslim-majority population. The Kashmir Valley is famous for its beautiful mountainous landscape, as well as Amarnath pilgrimage and Jammu's numerous shrines, particularly Shri Mata Vaishno Devi attracts tens of thousands of Hindu pilgrims every year. Ladakh is renowned for its remote mountain beauty and Buddhist culture. As a result of this religious scenario, three important traditional medical systems are practised in the region – Ayurveda in Jammu area, Unani in the Kashmir Valley, and Amchi in Ladakh area.

DOI: 10.4324/9781003329459-1

Socio-Economic Change

Kashmir had undergone various phases of historical and political development since the beginning of the 19th century, and therefore, socio-cultural economic and political conditions resulted in wide variation in the attitudes towards health and medical care systems in Kashmir.

The Arabs and the early Muslims attached great importance to care of the body, and therefore, cleanliness and health were acts of healing and piety. Sultan Zainul Abedin, the "Akbar of Kashmir" (1423–72 AD), ordered the translation of several medical works and patronized, among others, the great physicians Sribhatta and Mansur bin Ahmad. The Unani system of medicine together with spiritual medicine was the dominant system in the Kashmir Valley until 1865 when the Missionary doctor, Dr William Jackson Elmslie (1832–72), established an allopathic dispensary in Srinagar.

Some modern, allopathic physicians had visited Kashmir during the early 17th century, especially Dr F. Bernier (1625–88), a French doctor who was a personal physician to the Mughal Emperor Aurangzeb (1618–1707). However, Dr Elmslie was the first medical missionary who arrived in Kashmir in 1864. Subsequently, other medical missionaries, including the brothers Dr Arthur Neve and Dr Ernest Neve, spent several years in Kashmir during the later half of the 19th century and the early part of 20th century. The diffusion of modern medicine through medical missionaries led to a conflict with the local Hakims – practitioners of Unani medicine – who protested before the Maharaja of Kashmir not to grant favour to medical missionaries.

Poor economic conditions and unsanitary and unhygienic living conditions present an environment conducive to disease. Various infectious diseases and Kangri burn cancer (specific to the region) were widely prevalent in Kashmir. The report dated 10 December 1861 of the Deputy Commissioner on special duty at Kashmir to the Secretary to the Government of Punjab bears testimony to the poor economic conditions of the people in Kashmir. The report states "The people in Kashmir are wretchedly poor and in any other country their state would be almost one of starvation and famines" (Akhtar, 2011). Under such socio-economic circumstances, it was not possible for people to maintain a hygienic environment (Dev, 1983). Dr William Jackson Elmslie described in his letters the unhygienic practices of the people that result in the widespread incidence of skin diseases. In one letter, he wrote:

> But there are other things in Kashmir which most terribly detract from its pleasure as a place of residence. The dirt is beyond description. Who can tell what Kashmir smells are? Not the odour of roses, such as one has expected to fill the air; but oh! Such, that the dirtiest of London Courts is sweeter than the cleanest of Kashmir villages.
>
> (Elmslie, 1875)

Elmslie was a keen surgeon and performed many procedures under abysmal conditions but with successful results. This included the first lithotomy, for bladder stone, performed on 23 May 1866. He reported 30 cases of skin epithelioma and suggested its relationship with the use of Kangri, a clay fire pot used close to the skin to keep warm in winter. Elmslie died while crossing the mountains in 1872 (Mir and Mir, 2008).

Walter Lawrence (1857–1940), a Settlement Commissioner in Kashmir, also noted in 1895 that "the clothes of the villagers are simple and extremely poor in appearance" (Lawrence, 1895).

As stated, socio-economic and environmental conditions were conducive to sickness and occurrence of disease until the middle of 20th century (Dev, 1983). Most of the diseases such as waterborne, gastroenteritis, jaundice, and enteric diseases were prevalent. Cholera, respiratory diseases, tuberculosis, and even leprosy were common. Lawrence wrote:

> For people who can travel the valley offers a climate perhaps unsurpassed by that of any country, but in July and August, Srinagar, surrounded as it is by swamps, is apt to prove unhealthy and depressing. Speaking generally the valley may be said to be fairly free from disease. Malarial fevers, liver complaints, consumption and dysentery are rare, and typhoid fever is said to be unknown.... But certain diseases are unfortunately very common. In October and throughout the winter smallpox is very common and causes great mortality among children. Goitre is frequent and is especially common in villages where the drinking water comes from limestone or magnesian limestone rocks. But the great scourge of Kashmir is cholera, for when once this terrible epidemic enters the valley the mortality is very heavy.
>
> (Lawrence, 1895)

Gulzar Mufti's Book

Gulzar Mufti's book *Kashmir in Sickness and in Health*

> presents an insights of Kashmir's ailing political health since the beginning of Dogra rule more than 150 years ago, until the present time. The author describes various aspects of British involvement in Kashmir, describes the pioneering work of the UK missionaries in its social, educational and healthcare development, It describes the impact of political events in the international arena on Kashmir.
>
> (Mufti, 2013)

Arthur Neve's Work

As stated earlier, some physicians of modern medicine (allopathic) visited Kashmir during the early 17th century. However, Dr William Jackson Elmslie was the first medical missionary who arrived in Kashmir in 1864. Subsequently, a number of medical missionaries, including Neve-brothers – Dr Arthur Neve (1859–1919) and

Dr Ernest Neve (1861–1946) – spent several years in Kashmir during the later half of the 19th century and the early part of 20th century. The diffusion of modern medicine through medical missionaries led to a conflict with local Hakims – practitioners of Unani medicine, who protested before the Maharaja of Kashmir not to grant any favour to medical missionaries. Among them the contribution of Dr Arthur Neve was outstanding. He was a medical missionary and a meticulous geographical traveller (Akhtar, 2011).

Walter Lawrence Writes about Arthur Neve's Research Findings (Neve, 1910)

> I remember a paper written by Dr. Neve and published a few months ago. He these proved the existence of three parallel ranges in the far west of the system of the Karakoram. Beyond doubt two of them belong to the most gigantic systems on the Earth's Surface.
>
> (Lawrence, 1911)

Some of the countries he describes are the most magnificent in the world. The great range of Karakoram Mountains, the unique Indus Valley, the lofty Himalayan peak Nanga Parbat, and the twin peaks of Nun Kun are amongst the most important subjects he deals with. He explored new country, crossed passes not visited before by Europeans, climbed several times to over 20,000 feet, and the same time has always been ready to help all the natives he came in contact with medical advice and by performing such small operations, as were possible under the circumstances (Lawrence, 1913).

Sir Welter Lawrence, who had spent six years in Kashmir and published his famous book – *The Valley of Kashmir* in 1895, commented on the work of Arthur Neve (Neve, 1911)

> If I had had the privilege of listening to Dr. Neve before attempted to write my poor chapter on Physical History in "The Valley of Kashmir", I should have been able to write with much greater effect. From what I have seen myself, I can testify to the great accuracy of Dr. Neve's account. I think this is a good opportunity, before this great audience, for letting you know what the work of the Medical Mission in Kashmir is doing. I lived six years in that country, and know the road from Kashmir to Gilgit and Ladak. Wherever I went there was only one question. The people did not want to see me, but they wanted to know when Neve "sahib" was coming. Neve Sahib, who brought comfort and healing wherever he went. Working with very little help, working in a very small way against every hindrance, against the Brahman influence, the two Neves (Arthur Neve and E.F. Neve) have won everything to them, and now they have a grand hospital in Srinagar, and when the Neves are not going into the villages, the villagers are coming into the Neves'.
>
> (Lawrence, 1911)

Several writers have highlighted the medical and humanitarian role played by these missionary physicians and nurses since 1864 (Dev, 1983; Geographical Journal, 1913):

> Scottish missionary zeal and healthcare traditions, they were able to initiate, guide and influence the development of medical treatment facilities in the state of Kashmir. They continue to inspire and remind us of the core values of caring for the sick in deprived areas. At the site of the old Mission Hospital in Srinagar now stands the largest Chest Disease Hospital in the valley, a monument to the great medical missionary pioneers who laid down their lives in the service of the Kashmiri people.
>
> (Mir and Mir, 2008)

Current Health Scenario

The findings and analysis of contemporary health scenario in the region reveal that among 31% of the hypertensive, the majority were males from urban areas living in nuclear families, having high educational qualifications (Ph.D.) and belonged to high-income group. Majority had the onset of the hypertension in middle age and was not having any family history of high blood pressure. In addition, they were having other diseases/disorders (Shakeel and Irshad, 2017).

Mufti has rightly asserted that "With urbanization and increasing prosperity, prevalence of hypertension is on rise, about 60% population of Kashmir is hypertensive and lifestyle is an important risk factor for it. Hypertension thus becomes an increasingly important clinical problem" (Mufti, 2013).

Several researchers agreed that a rising trend in ischemic heart disease (IHD) is associated with the risk factors like smoking, diabetes, obesity, sedentary lifestyle, and pollution in both urban and rural areas.

Besides the hypertension, cancer is on rise in Kashmir (Rather and Akhtar, 2000). Earlier in the 19th century and early 20th century, Kangri burn cancer was predominant, but now incidence of all forms of carcinomas is on rise (Akhtar, 1992).

According to ICMR, in 2017 during seven months period, 16,480 new cases of cancer were registered in Jammu & Kashmir. Greater Kashmir reported that 87% rise in cancer cases was recorded in Jammu & Kashmir.

Based on the ICMR Report on Jammu & Kashmir – Disease Burden Profile during 1990 and 2016, death and disability scenario changed drastically during the period. In 1990, diarrhoeal diseases ranked first in Jammu & Kashmir, which shifted to eighth rank in 2016. IHD ranked first followed by chronic obstructive pulmonary disease (COPD) and road injuries in 2016. In 1990, COPD was at fifth place in death and disability ranking.

Another study that analysed the cancer data of Jammu suggests that leading sites of cancer in males were lung (9.6%), myeloid leukaemia (8.3%), prostate (6.8%), mouth (6.1%), and gall bladder (6.0%), while in females, breast (35.7%), cervix (19.1%), oesophagus (5.1%), myeloid leukaemia (4.7%), and gall bladder (3.9%) were the major cancer sites. The author asserts that population-based epidemiological studies are required to find out the disease burden and its cause in this region (Singh et al., 2016). A geographical study reveals the distribution of cancer in Kashmir (Rather and Akhtar, 2000).

Vector-borne diseases such a malaria and dengue are major health problems in the Jammu region. In Jammu region, rising cases of dengue are causing panic. Most cases were reported from Jammu district followed by Sambha, Kathua, Udhampur, Doda, and Raouri districts. (J.V, 2017).

In a study conducted in Ladakh region, Rehman and colleagues assert that the most common ailment prevalent in the Ladakh division is arthritis followed by peptic disorders (Mayer and Akhtar, 2008). However, most diseases like diabetes, malaria, TB, carcinomas, heart ailments, mental disorders, and other cardiovascular diseases are less prevalent in the region. The study also showed that the physical environment and socio-economic conditions were associated with the prevalence of these diseases (Rehman et al., 2004).

Climate Change, Mountainous Regions, and Diseases

A new study finds that disease outbreaks are increasing at higher altitudes around Mount Kenya, thought to be as a result of global warming (Akhtar, 2007; Williams, 2010).

In 2010, India's Ministry of Environment and Forests released its first assessment report of the projected impacts of climate change by the 2030s on key sectors in the country's climate-sensitive regions, including the Himalayas and the northeast. The report was among the first to flag a study by scientists at the NIMR that established a nexus between climate change and health. These studies suggest that climate change will significantly increase both the intensity and geographic spread of malaria and other vector-borne diseases in areas that have been largely insulated from them in the past.

Rising temperatures due to climate change will lead to the altitudinal occurrence of malaria in several hilly regions, and vector-borne diseases like malaria and dengue may occur in the Himalayan region, including Mizoram, Manipur, Arunachal, Uttarakhand, Himachal Pradesh, and some parts of Jammu & Kashmir (Akhtar, 2007). Experts suggest that there is urgent need to strengthen the region's health infrastructure and increase people's awareness about the impact of climate change on the outbreaks of vector-borne diseases (GK, 2015).

Thus, socio-economic factors including urbanization, globalization industrialization, and economic prosperity in general have resulted in partial lack of epidemiological transition in Jammu & Kashmir.

From early 2020, particularly from March 2020, the whole world, including Jammu & Kashmir and Ladakh, started experiencing the impacts on health caused by coronavirus (COVID-19). COVID-19 has caused global health and economic disaster that has not been experienced in recent time. It caused havoc to all geographical regions including the Himalayas. There is need to study the spatial spread of COVID-19 – identify high- and low-intensity regions and pinpoint regions impacted by Delta variant, and discuss various socio-economic and physical determinants of the distribution pattern of COVID-19 in all geographical regions in Jammu & Kashmir and Ladakh. COVID-19 cases are still rising in both Jammu & Kashmir and Ladakh. As of 14 August 2022, there were 473,782 cases with 4777 deaths in Jammu & Kashmir and 28,271 cases with 228 deaths in Ladakh.

References

Akhtar, R. (1992) Kangri: Traditional Personalized Central Heating in the Valley of Kashmir, *Focus*, Journal of the American Geographical Society, Fall, Vol. 42, 2, pp. 25–27.

Akhtar, R. (2007) Health and Climate in Kashmir, *Tiempo: A Bulletin on Climate and Development*, International Institute for Environment and Development (IIED), London, Issue, 63, April, pp. 19–21.

Akhtar, R. (2011) Arthur Neve (1859–1919) and a Mission Hospital in Srinagar, Kashmir, *Journal of Medical Biography (Royal Society of Medicine*, London), Vol. 19, pp. 177–181.

Dev, S. (1983) *Natural Calamities of Jammu & Kashmir*, Ariana Publishing House, New Delhi.

Elmslie, W. J. (1875) *Seedtime in Kashmir: A Memoir of William Jackson Elmslie*, Thomson, W Burns, James Nisbet, London.

Geographical Journal (1913) Thirty Years in Kashmir, Review of the book of Arthur Neve, *Geographical Journal*, Vol. 43, 4, pp. 429–443.

Greater Kashmir (2015) Climate Change to Make JK Vulnerable to Malaria, *Dengue*, August 20.

Hunter, John (1974)The Challenge of Medical Geography, In *Hunter, J* (ed.) The Geography of Health and Disease, University of North Carolina, Chapel Hill (Dept. of Geography, Studies in Geography No. 6, p. 15).

J.V. (2017) In Jammu Rising Cases of Dengue Case Panic, Myth and Reality. http: //www.jammuvirasat.com/2017/10/21

Kirk, W. and Akhtar, R. (1990) Jammu and Kashmir State of India, *Encyclopaedia Britannica*, Last updated January 31, 2019 (https://www.britannica.com/place/Jammu-and-Kashmir)

Lawrence, W. R. (1895) *The Valley of Kashmir*, Henry Frowde, London.

Lawrence, W. R. (1911) Comments on the Presentation Entitled-Journeys in the Himalayas Delivered by Arthur Neve, at the Meeting of the Royal Geographical Society, London.

Lawrence, W. R. (1913) Comments on the Presentation Entitled-The Ranges of the Karakoram, Delivered by Arthur Neve, at the Meeting of the Royal Geographical Society, London.

Mayer, I. A. and Akhtar, R. (2008) Disease Ecology of Ladakh Himalayan Region, *Annals National Association of Geographers*, India, Vol. 28,No. 2, pp. 40–48.

Mir, N. A. and Mir, V. C. (2008) Inspirational People and Care for the Deprived: Medical Missionaries in Kashmir, *Journal of the Royal College of Physicians of Edinburgh*, Vol. 38, pp. 85–88. https://www.ncbi.nlm.nih.gov/pubmed/19069044

Mufti, Gulzar (2013) Kashmir in Sickness and in Health, Partridge Publishing, Gurgaon.

Neve, A. (1911) Journeys in the Himalayas, *Geographical Journal*, Vol. 38, pp. 345–362.

Neve, A. (2010) The Ranges of the Karakoram, *Geographical Journal*, Vol. 36, pp. 571–577.

Rather, G. M. and Akhtar, R. (2000) Cancer Mortality in Kashmir, *Indian Geographical Journal*, Vol. 75, No. 2, pp. 131–140.

Rehman, S.U., Faqtoo, A. Q. and Ahmad, B. (2004) A Study on "Disease Prevalence" in Ladakh, Jammu and Kashmir, October. (https://www.researchgate.net/publication/291856125_A_study_on_disease_prevalence_in_Ladakh_Jammu_and_Kashmir)

Shakeel, S. and Irshad, N, (2017) Lifestyle Patterns and the Prevalence of Hypertension Among the Teachers of Kashmir University (Age 35 To 60 Yrs), *International Journal of Home Science*, Vol. 3, No. 1, pp. 150–154.

Singh, G., Mahajan, S. and Suri, A. (2016) Pattern of Cancers In Jammu Region, *JK Science*, Vol. 18, No. 3. www.jkscience.org.

Williams, N. (2010) Malaria Climbs the Mountain, *Current Biology*, Vol. 20 No. 2, p. R3.

2 Environment and Cholera in Kashmir during 19th Century

Rais Akhtar

Introduction

Historical Medical Geography is a fascinating area of research which highlights the pattern of environmental conditions, people's adjustment with the environment and the occurrence of health problems as a result of man's interaction with the environment.

The outbreak of cholera and its diffusion in Kashmir in the 19th century was more or less considered a local epidemic with its origin in Kashmir and in the neighbouring Punjab. However, the examination of data and other source material on the outbreak and diffusion of cholera at the global scale in general and India in particular, reveal that cholera smouldering in an environmental hearth in the Gangetic delta of Bengal, erupted in four epidemic waves in the 19th century, which caused sudden and widespread mortality throughout the world. North America was stricken three times. In Europe the spectre of advancing death, fuelled by word of mouth reports, created an advance wave of panic, heightened by total public ignorance of cholera's aetiology. The eventual discovery that cholera is a water-borne infection was made during the cholera epidemic in London in 1854 by Dr. John Snow. According to J. M. Hunter:

> With the widespread adoption of steam boats and the improvement of shipping lanes, global waves of the so- called Indian cholera spread even more quickly. The North American pandemics of 1832, 1848, and 1866 variously entered the seaports of Quebec, New York, and New Orleans and spread along river valleys and waterways. The disease passed through Buffalo, the Great lakes, Detroit River, Chicago and principally along the Mississippi River cholera was even carried west in the California gold rush . Later, the geographical pattern of spread of the disease in North America began to change as towns and railroads were constructed.
>
> (Hunter, 1988)

A study carried out by G. F. Pyle showed that the 1866 epidemic was strongly influenced by the urban hierarchy, and was also related to distance from the eastern seaboard. Thus a regular, geographically predictable relationship was observable in the spread of the disease (Pyle, 1979).

DOI: 10.4324/9781003329459-2

The above discussion throws light on two important aspects:

1 To study the cholera ecology in Kashmir;
2 To test the hypothesis that cholera was not indigenous to Kashmir and was part of the process of cholera outbreak and diffusion globally; since cholera occurred in Kashmir, and northern Indian sub-continent, Europe including the United Kingdom and the USA more or less at the same time.

Geo-Ecology of Cholera in Kashmir

Physical Factor

The valley of Kashmir with its clayey loamy soils, nestles in the young folded mountain ranges of the Himalayas. The water in the Kashmir valley is an extraordinary paradox. Considering the total run-off, area of water-bodies and length of water courses, the valley has no match in the Himalayas. In fact its water features are the principal components in its scenic beauty. The geomorphic character of the valley is however, such that the distribution of water resources is extremely uneven – a situation which renders vast stretches of land totally or partially out of use either due to excess of water or its deficiency. Water is most plentiful in the low lying parts of the valley, which remain literally deluged, while the adjoining *Karewa* unplands suffer from aridity imposed by its chronic deficiency. The consequence is that the valley presents the anomalous case of scarcity in the midst of plenty. The rivers carry large volumes of water which they cannot possibly contain as their channels get increasingly choked with silt, making floods a recurrent phenomenon with disastrous consequences on agriculture and on health simultaneously. It was this phenomenon of water scarcity in the midst of plenty as a result of particular geomorphological pattern that used to cause famine in the valley (Raza et al., 1978). When famine is combined with cholera as was the case during 1877–79 it caused havoc among the people of Kashmir. Another aspect of geomorphic characteristics of the valley is the pattern of soil distribution. Writing in early 1890s, Walter Lawrence who was then Settlement Commissioner noted, "it is a curious fact, which I noticed in my tours, that villages on the *Karewa* plateaus seemed free from cholera, and that the disease was most rampant in the alluvial parts of the valley" (Lawrence, 1895). An examination of geo-ecology of cholera reveals among other factors that, apart from man (as both pathogens and geogens), there are environmental factors which facilitate the survival and spread of cholera. Such geographical factors include: annual temperatures of around 37°C and prolonged dry spells. Moisture is another important factor which favours the outbreaks of cholera. Cholera vibrios thrive best in warm alkaline media (with pH value between 9 and 9.6) while acidic media inhibit their growth and survival. The *Karewa* soils,[1] however, are normal with a pH value of around 7.5 and therefore not acidic. However, the *Karewa* soils are devoid of vegetal cover and are deficient in organic matter. The moisture-retaining capacity of the soil is poor as the upper layer has a high sand content (Raza et al., 1978). The characteristics of soils may explain the statement of Walter Lawrence regarding the absence of cholera in the villages on the *Karewa* soils of Kashmir valley.[2]

Poor Socio-Economic and Insanitary Conditions

A report dated 10 December 1861 from the Deputy Commissioner on special Duty at Kashmir, to Secretary to Government of Punjab, is testmony of the poor economic conditions of the people in Kashmir. The report states, "The people in Cashmir are wretchedly poor, and in any other country their state would be almost one of starvation and famines ...." (Report, 1861). Under such socio-economic circumstances, people were not able to maintain a hygienic environment. Dr. William Elmslie, who arrived in Srinagar as a missionary doctor in 1865, described in his letters the unhygienic practices of the people which resulted in the widespread incidence of skin diseases. He wrote:

> But there are other things in Kashmir which most terribly detract from its pleasure as a place of residence. The dirt is beyond description. Who can tell what Kashmir smells are? Not the odour of roses, such as one has expected to fill the air; but oh! such, that the dirtiest of London courts is sweeter than the cleanest of Kashmir villages. The clothes, too, of the people are filthy; not that the filth shows much, for all their garments are of grey wool, which is a most perfect concealer of dirt; but not a few of their diseases are the result of their uncleanliness, and how often I have almost shrunk away from them, as, in my dispensary, while I have been examining a patient. I have seen the lice crawling on his clothes and his fleas skipping over to me. Of course, if you can avoid all intercourse with the natives, then dirt is not much a continual source of annoyance, but to us it was a daily trouble.
>
> (Elmslie, 1875)

Similar views regarding insanitary conditions, in Srinagar were expressed by Irene Petrie who was very perturbed when she wrote:

> Srinagar, with its water-ways, palaces, bridges and graceful fair skinned inhabitants, suggests Venice, though Venice much dilapidated. But from Venice there are no such views as one may see on a clear autumn day. It is perhaps the dirtiest city in the world, and most of the houses look as if they could not survive the next flood or earthquake.
>
> (Wilson, 1900)

Irene Petrie pleaded that people should be taught practically that cleanliness is next to godliness.

Most writers on Kashmir agreed to the fact that cholera outbreak was the result of "poor sanitary conditions." The centre and nursery of cholera in Kashmir is the foul and squalid capital Srinagar, but if it is once established there it soon spreads to the dirty towns and villages (Lawrence, 1895, p. 219). Besides crowding was another serious problem. Nearly 1,18,960 people live in 22,448 houses with a density of five persons per house. The houses were:

> built irregularly and without any method, on narrow tortuous paths. Ventilation in the town is therefore very imperfect. Few houses have latrines, and small lanes and alleys are used as such....There is no drainage. Slush, filth and ordure are washed by storm water into the river and Nalla Mar which supply the city with drinking water. On account of absence of snow in winter and rains in spring, and river was dry and low and the bed of the Nalla Mar canal was converted into a string of cesspools. People were immersed in a polluted atmosphere caused by the products of putrefactive and fermentable water accumulated in houses and numerous narrow lanes, passes, nooks and crevices which intersect the town. This produced an epidemic constitution in the people fitted for the reception and fostering of cholera-germs.
>
> (Lawrence, 1895, p. 36)

Riverine Environment

The riverine environment, too, contributes to a greater extent towards cholera outbreak. The riverine environment was significant, in Kashmir in the 19th century. Most of the localities lined the Jhelum river for a length of about 5 kilometres, and the town extended less than a kilometre, with Habba Kadal locality as most densely populated area, with the river as the main source of drinking water. Drew compared the river Jehlum with the Thames, and wrote a vivid account of the layout of Srinagar:

> It is (Jhelum) the chief artery of traffic; it is of much more importance as a thorough fare than any of the streets; indeed there are one or two streets, and those but short ones, that have anything like a continuous traffic, while the river is always with boats.
>
> (Drew, 1875)

About the riverine environment Drew further noted:

> A few canals traverse the interior of the town; one of them is wide, and is overlooked by some of the best of the houses. One is narrow, passing through some of the poorest parts, low dwellings crowd on it that, albeit they are well peopled, seem to be on the point of falling....A third canal leads from the upper part of the city to the gate of the lake, and shows along its winding course groves of plane-trees on the banks that make a beautiful combination with the smooth waters at their feet and the mountains that rise behind them.
>
> (Drew, 1875, p. 185)

Dr. William Elmslie, who arrived in Kashmir as a medical missionary, wrote about the outbreak of cholera in this riverine locality of Habba Kadal in 1867, "On 20th June, cholera has broken out in the city. The worst was in Habba Kadal" (Elmslie, 1875). Superstitious beliefs were a hindrance in controlling the spread of cholera. Writing about 1872 cholera outbreak, C. L. Tyndale Biscoe, who was an educational missionary, noted:

> The flood was followed by an epidemic of cholera, in which it was difficult for the English medical missionaries to help as they wished, because the people, believing their sickness to be judgement from God, trusted chiefly to charms to be swallowed in water. These were pieces of paper, written by the priests and bought from them on which the name of ALLAH, or of some Mohammedan saints, or, in case of Hindus, of some god or goddess, was inscribed. The people continued drinking infected water all the same, and consequently cholera stalked through the land unchecked, claiming its thousands of victims.
>
> (Biscoe, 1921)

Movement of Polulation Opening of Jhelum Route, and Troop Movement

With the commencement of Sikh rule in 1819, contact between Kashmir and Punjab began especially when Kashmir faced famine and there was acute shortage of food in the valley leading to a large-scale migration to Punjab which contained several foci of cholera in north-western India. According to Lawrence, "a number of Kashmiris fled to Punjab as a result of famine in 1831. Sikh governor of Kashmir imported grains and eggs from Punjab and restored some measure of prosperity to the villagers who had lost their grain seed and fowls in the awful famine" (Lawrence, 1895, p. 200). However, the Sikh governor, considered sympathetic to Kashmiris, was murdered and the turmoil brought about by Bombas followed by raids by both Bombas and Kukas, Maharaja Gulab Singh, the Dogra[3] ruler of Jammu region took over the rule of Kashmir in 1846. This has further intensified the contacts between Kashmir with the cholera-prone region of Jammu, south of Kashmir. Travel between Srinagar and Jammu, especially the Maharaja's camp, brought cholera from Jammu. Dr A. Mitra, who wrote *Medical and Surgical Practices in Kashmir* (1890), opined:

> Cholera is not endemic in Kashmir, but visits it often in virulent epidemic form. The last epidemic as in 1880 during which nearly 10,000 persons died from cholera. The disease travelled with the Maharaja's camp from Jummoo to Srinagar, where the first case was observed on 6th April. Day by day the number of cases began to increase till on the 15th of May 250 persons were attacked in Srinagar alone.
>
> (Mitra, 1889)

Describing the horror of 1888 cholera outbreak, Mitra noted:

> Its ravages extended to all parts of the valley. The scene of death and desolation during the summer of 1888 was one that will not soon pass away from living memory. The country was gay, magnificent arrangements were being made for a fitting reception of the Earl Dufferin. The Maharaja and suite arrived earlier than usual to loyally welcome the Viceroy. Suddenly the disease epidemic came, the Viceregal visit was abandoned and everybody became panic-stricken. The Maharaja and suite and the Resident remained in Srinagar but no doubt in great trepidation.
>
> (Mitra, 1889, p. 11)

The intensity of cholera occurrence increased in Kashmir from around 1890 when the road between Punjab and Baramulla in Kashmir was thrown open. Lawrence wrote in grave fear, "Now that Srinagar is joined to India by a road, there is a two-fold necessity for sanitary reform. For if cholera becomes endemic the Punjab and the great military cantonment of Rawalpindi will always be threatened, while on the other hand, the occurrence of an epidemic in India is sure to be followed by cholera in Kashmir, for cholera like trade, travels by roads. Before the road from Baramulla to the Punjab was opened cholera might occur in India while Kashmir was healthy, and whereas there were twelve epidemics of cholera in the Punjab between 1867 and 1890, there were only five outbreaks in Kashmir during the same period. Now it is almost certain that if cholera reaches the Punjab it will find its way along the crowded road and the narrow valley to Srinagar" (Lawrence, 1895, p. 219).

It is interesting to note that a report on 1862 Punjab cholera prepared by J.B. Scriven made similar observations. The report noted, "The route of the cholera was particularly well marked in the year 1861, in which it was regularly traced along the grand trunk road, from Delhi to Lahore. This year (1862), likewise it appears to have transmitted in two lines, one on the east and the other on the west of Punjab leaving the immediate country free" (Report, 1861).

Dr. Elmslie's Account

Dr. Elmslie narrated his experiences of cholera outbreak of 1867. He put forward his theory that the Maharaja's sepoys who went to Hardwar (a religious centre in northern India), brought cholera infection:

> just as the hostility and opposition of the local government of Kashmir had reached a climax cholera broke out amongst his Highness's troops. It appears that some sepoys who had got leave to go and wash in the Ganges at Hard-war, had returned to their regiment at Srinagar, and brought the seeds of cholera with them. At any rate, those sepoys had scarcely arrived, when that awful pestilence broke out and began to carry off many. Everything was done

to prevent the spread of the disease, but it at last invaded the city: and the British Resident deemed it necessary for the safety of the European visitors, to institute a *Cordon Sanitaire* round the European quarters. As my dispensary is situated there, I was compelled to put a stop to my work for a time.... The poor people of the city are sadly neglected, even by those who ought to take some care of them.

(Elmslie, 1875, pp. 195–196)

As Part of Pandemic

Having studied the geoecology of cholera in Kashmir in the 19th century, it is worthwhile to throw light on the pandemics of cholera. F. G. Clemow in his book: *The Geography of Disease* (1903), opined that "the pandemics of cholera have always started from India" (Clemow, 1903). Cholera, smouldering in an environmental heart in the lower Ganges valley, erupted in four epidemic waves in the 19th century, which caused sudden and widespread mortality throughout the world. North America was stricken three times. In Europe the spectre of advancing death, fuelled by word of mouth reports, created an advance wave of panic, heightened by total public ignorance of cholera' aetiology (Hunter, 1988, p. 117).

Diffusion of Cholera from Bengal to Kashmir

F. N. Macnamara, whose book on *Climate and Medical Topography*, was published in 1880, provided the most scientific account on cholera diffusion in northern India (Macnamara, 1880). Based on the analysis of data obtained from the book by Macnamara, the study reveals that cholera outbreaks occur in Bengal during February and March, move through northern India to Punjab and North-West Provinces, where its outbreaks are generally reported in the months of August and September. Because of Kashmir's connection with Punjab, cholera moved to the valley and occur normally during winter months, November–January as mentioned earlier. The analysis of data for the period 1865–76 for Bengal, and 1870–76 for Punjab and North West Provinces reveals the following pattern of cholera outbreak in northern part of Indian sub-continent (Table 2.1):

Table 2.1 Cholera out-break pattern in Bengal, Punjab and Northwest Provinces (19th Century) (Macnamara, 1880, p. 112)

Province	Total cases (average annual)	Outbreak occurred
Bengal (1865–76)	4,657	February–March
Punjab(1870–76)	3,019	August–September
N.W. Provinces (1870–76)	21,960	August–September
Kashmir (1870–76)	Several Thousands	December–January
(1888)	10,000	May–August
(1892)	18,000	May–August

It is evident from the Table 2.1 that the intensity of cholera generally increased as it moved westwards fron Bengal to N.W. Provinces. In Kashmir, cholera outbreaks were reported both in summer as well as winter. Cholera outbreak in Bengal in February or a little later during slight spring rains, moved to northern India and became epidemic till monsoon winds bring humidity, if not actual rainfalls (Macnamara, 1880, p. 112).

It is interesting to note that cholera broke out in Kashmir during both a snow-free winter as well as a winter period which experienced heavy snowfall. In the *Gazetteer of Kashmir and Ladak*, Bates stated that "in 1857 cholera struck to the valley, strange to say, throughout the winter, when the snow was up to a man's neck" (Bates, 1890). Contrary to this Lawrence wrote in 1895 that:

> on account of absence of snow in winter and rains in spring, the river was dry and low and the bed of the Nala Mar Canal was converted into a string of cesspools....This produced an epidemic constitution in the people fitted for the reception and fostering of cholera-germs.
>
> (Lawrence, 1895, p. 36)

Thus it may be concluded that the outbreak of cholera in Kashmir was result of the pandemic wave and not the local environmental conditions, as cholera broke out both during snow-free and heavy snow periods.

Cholera Ecology Model

Table 2.2 provides an insight of the ecological conditions which favour cholera outbreaks in Kashmir. It reveals that not only the physical factors but socio-economic and cultural factors especially mobility of population between Kashmir and cholera foci region of Punjab play a significant role in the outbreak and the diffusion of cholera.

Table 2.3 depicts the pattern of cholera diffusion model based on the movement of cholera from Bengal to Punjab and North-West Provinces. From Punjab, one wave of cholera diffusion moved to the valley of Kashmir, mainly via the Jhelum

Table 2.2 Cholera ecology model in Kashmir (1850–1900 AD)

a Physical factors→	Rains + Estuarine environment + prolonged dry spells + Sluggishly flowing water + alkaline soils
b Socio-economic and cultural factors →	Unhygienic living conditions + Contaminated drinking water + Defecation in the lanes + Growing contact with Punjab (migration) at the Commencement of Sikh rule followed by Dogra rule + Famine + Incoming of troops from Cholera Foci Areas

Source: Based on data from M.K. Dutta, 1973.

Table 2.3 Diffusion model of cholera in Kashmir (during 19th century)

Calcutta (Ganga Delta)	→	Delhi	→	Jhelum
Jullundhar	→	Jhelum	→	Peshawar
Lahore	→	Srinagar		
	←	Srinagar		
Lahore	→	Ambala		
Ambala	→	Jumbo	→	Srinagar

Source: Based on data from Bellew, 1885.

Table 2.4 Outbreak of Cholera during 1845–67*

Kashmir	Murree	Punjab	N.W. Provinces
1845	1858	1861	1862
1857–58			
1867			
Northern India		Central India	Rajasthan/Rajpootana
1867		1864,	1856
		1866	1866
		1869	1869
UK		USA	Nepal
1854,		1848–50	1857–58
1866–67		1866	

* Based on the table produced by Bellew, 1867 and 1879 were years of severe outbreak of cholera in Kashmir and in the Indian sub-continent. Similarly 1874 and 1877 were cholera free years in both these regions (Bellow, 1885).

route and the other wave of cholera reached Peshawar (Figure 2.1). Besides, diffusion of cholera into Kashmir, also took place via the Central Asian route, possibly via Kabul. It is pertinent to mention that there were intense trade links between Kashmir and Afghanistan; a number of Afghanis were in the service of the Maharaja of Kashmir (Kabul, 1882, 1885, National Archives).

Table 2.4 also reveals that cholera outbreaks in Kashmir were not isolated events but part of the pandemic wave spread throughout the world in general and the Indian sub-continent in particular. For example cholera occurred in Kashmir in 1845, and three years later it broke out in the USA and continued until 1850. Outbreak of cholera was reported in London in 1854, in Kashmir 1857–58; in Murree hill station in Pakistan in 1858, and in Nepal during 1857–58:

Rodenwaldt and Jusatz have published a World Atlas of Epidemic Diseases, in three volumes during 1952–61, which includes a map of worldwide diffusion of cholera (Figure 2.2) during the period 1863–68. The diffusion pattern reveals that cholera originated from Bengal and had diffusion through central India into Africa during 1863 and 1864. This could be linked with the migration of Indian labour force to British colonies in Africa in the 19th century. From Africa it reached the Middle East in 1865 and towards Central Asian countries where cholera outbreak continued to occur upto 1872. In view of the intense contacts between Central

Asian countries and Kashmir, cholera arrived in Kashmir in 1867. According to G. F. Pyle (1979), the disease also:

> diffused southward through the Malay Peninsula and into the East Indies by 1865. Another track was into China. During the period 1865–67 the cholera spread into Europe and Russia, as well as down the west Africa coast. By 1866, the pandemic also had reached the New World (Figure 2.3). While entry into South America is not well known, Rodenwaldt's reconstruction suggests that the disease spread up the Rio La Plata and into the interior of the southern part of South America. Cholera spread rapidly to the Antilles in 1865 and reached Canada and the United States by 1866.

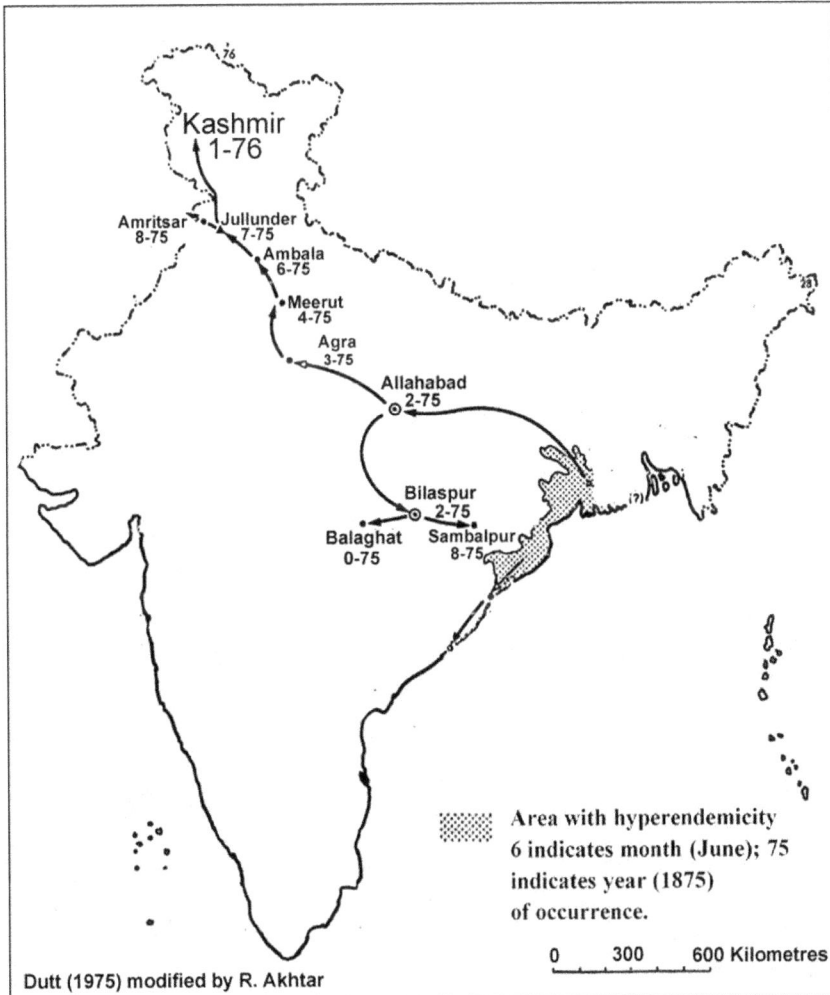

Figure 2.1 Spatial Diffusion of Cholera in Northern India (1875–77).

Thus it is evident that cholera outbreak of 1867 in Kashmir was a part of a pandemic wave.

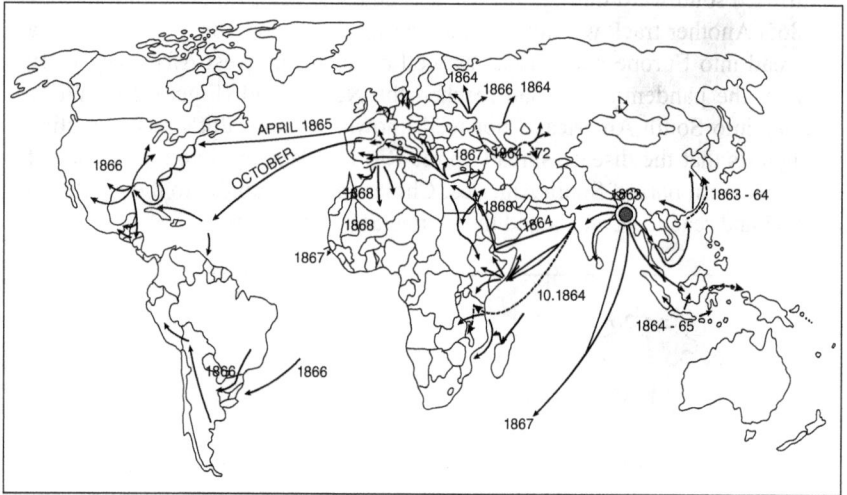

Figure 2.2 Worldwide diffusion of cholera 1863–68.
Source: G.F. Pyle (1979).

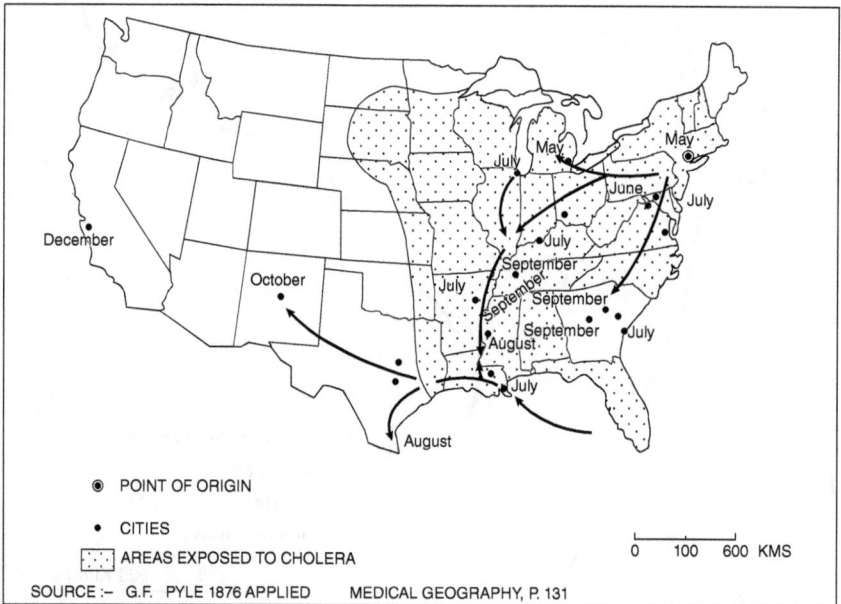

Figure 2.3 Diffusion of Cholera in the US 1866.
Source: G.F. Pyle (1979).

Insanitary Conditions

As regards the second hypothesis on insanitary conditions related to cholera outbreak, the overview of the literature on insanitary conditions in some European cities suggests that Kashmir was not exceptional, and such conditions prevail in several cities of Europe where towns have a reputation for being more unsanitary than rural areas (Jeffery, 1988). Following two examples are relevant:

Around the middle of 19th century,

Glasgow was possibly the filthiest and unhealthiest of all the towns of Britain at this period...Immigration into the city had occurred on a scale without precedent and accommodation provided to meet it was hopelessly inadequate.... Nor was there an effective system of refuse removal.

(Howe, 1972)

About the city of Exeter in England, Dr. Shapter wrote:

The adequate water supply combined with deficiency of drainage, is so self sufficient evidence, ... dwellings occupied by from five to fifteen families hudled together in dirty rooms with every offensive accompaniment; slaughter houses in the Butcher Row: with their put rid heaps of offal; of pigs in large numbers kept throughout the city.....poultry kept in confined cellars and outhouses, of dung heaps everywhere....

(Howe, 1972, pp. 181–182).

The above quotations are comparable to the account of insanitary conditions in Kashmir as described by Irene Petrie and Walter Lawrence. Thus, insanitary conditions prevailing during the latter half of 19th century were not exceptional in the context of Kashmir only, and there was hardly any sanitary improvement in Kashmir until 1892, when the severest outbreak of cholera occurred. The question arises as how to explain the outbreak of cholera in certain years. Mobility of population between Kashmir and Punjab and later with Jammu region could be the main source of infection and subsequent diffusion of cholera in Kashmir. It is also evident from the above discussion that cholera outbreak occurred simultaneously in several other towns and cities of Indian sub-continent as well as in other parts of the world including Europe and the USA.

Meteorological Evidences and Outbreak of Cholera

Regarding the sudden outbreak of cholera in Kashmir, the opinion of J. Fayrer, a medical climatologist needs consideration. Fayrer who wrote books on, *The Natural History and Epidemiology* (1888), and *Preservation of Health in India* (1894) was a prolific writer on medical climatology, had worked in India with the Indian Medical Service. Fayrer preferred atmospheric theory against water borne theory (then generally accepted by the British Medical Professionals a result of Dr. John Snow's work on Cholera in London in 1854). Fayrer noted that the suddenness and virulence of certain outbreaks of cholera are remarkable, and seem to point to some factor apart from contagion or local insanitary conditions. Rather, the evidence seemed to point to changing meteorological conditions:

> At Kurrachee, in 1866....there was a sudden change in the atmosphere, the wind veered from south-west to north-east, and a thick lurid cloud darkened the air. Later in the evening cholera appeared in thirteen corps of the troops stationed there.
>
> (Fayrer, 1888)

Karachi was also affected by sudden cholera outbreak two decades earlier. According to Sanitary Report of 1869. Cholera broke out in Karachi in 1846, and resulted in the death of 438 European soldiers out of 3,345, and 199 Native soldiers out of 3,045. The Sanitary Report states:

> It was in the evening of 14th June that Karrachi was struck. The epidemic occurred before Dr. Snow's theory was announced, and it is interesting to observe that among the many hypotheses that were advanced to explain its suddenness, its virulence, and its very partial distribution, no mention whatever is made of the state of water supply. No one appears to entertained the smallest suspicion that the virus could have been disseminated through that medium.
>
> (Report on Sanitary Administration, 1876)

It would not be out of context to mention that it seems Dr. John Snow's theory on cholera was quite well known by medical personnel working in Jammu, Kashmir and Ladakh regions during later half of 19th century. Caley, a British surgeon who worked at Leh dispensary of Jammu & Kashmir published a paper in 1868 in *Indian Medical Gazattee,* referred cholera as poison of the disease. It must be mentioned here that in 1849 John Snow published a small pamphlet entitled, "On the mode of communication of cholera", in which he proposed that "cholera poison" reproduced in the human body and was spread through the contamination of food and water. Thus it seems Caley was aware of the John Snow's work.

In the absence of detailed meteorological data, it is not possible to test the theory of meteorological explanation in the outbreak of cholera in Kashmir. However, two statements, cited earlier that of Dr. Elmslie and Dr. Mitra do give some idea of sudden atmospheric change in relation to cholera outbreak. On an earlier occasion in 1862, Maharaja was held up in Jammu and could not move to Srinagar as a result of cholera outbreak. "In May-June cholera raged with such virulence in Jummoo, that the Maharaja delayed his departure for Cashmere till it abated.... at the end of July – the season of periodical rains" (Elmslie, 1875, pp. 195–196).

During the 1867 outbreak, it seems rains were deficient until 2 August, when Dr. Elmslie wrote that the city of Srinagar experienced a great storm of wind and rain for about an hour and prayed, "May good result from this." It rained heavily in the coming few weeks, and according to Dr. Elmslie, the cholera seems to lie on the wane (Elmslie, 1875, pp. 195–196).

However, the analysis of temperature and rainfall data for 1892, the year of the severest outbreak of cholera, reveals that spring rains failed in that year, but the summer rains were in excess, as a result of very high mean maximum temperature during the months of July, August and September 1892. According to annual Administration Report, drier weather than usual during April to May followed by torrential rain, could be the reason for the severe outbreak of cholera in 1892.

Impact of Sanitary Measures

After the 1892 cholera outbreak in Kashmir, efforts were made towards improvement in sanitary conditions. These have brought desired results. According to the Annual Administration Report of 1895–96, the sanitation conditions in the city of Srinagar have improved considerably:

> Roads and drains have been made, a supply of pure water has been started and conservancy is systematically and methodically done. Dr. A. Mitra who was also Administrator, Srinagar Municipality during the 1892 cholera outbreak, says, "These measures of public sanitation are having their influence on the habits of the people, and thus the cause of both public and private hygiene is improving with rapid strides. The opprobrium now resting on Srinagar, as a filthy city, and on its inhabitants, as a filthy people, will, I confidently hope, be a thing of the past at no distant date. Never, perhaps in the history of sanitation so quickly and so effectively have sanitary improvements been done as in Srinagar, in spite of financial and other local difficulties.
>
> (J&K Govt, 1892)

Describing the horror of 1892 cholera outbreak, Rev. Tyndale Biscoe who arrived in Kashmir in 1890 and started a missionary school, commended the assistance of school boys who rendered valuable services. He wrote:

> I was very much gratified that at the commencement of the cholera about six boys came and offered their services to help look after the sick. Now the Kashmiris were in deadly terror of it, and one of these boys who offered to help died of it (such a dear fellow).
>
> (Biscoe, 1892)

Dr. E.F. Neve who was missionary doctor in the valley discussed the religious-cultural aspect of people's behaviour at the time of the 1892 cholera outbreak. People and faith in the power of shrines. According to Neve:

> And even the common people placed the most implicit trust in the protecting power of the shrines. In village after village, where the cholera did not come,

the people ascribe their escape to the influence of the village Ziarat (shrines) and the offerings which they make. The love of shrines and relics is deeply imbued in the heart of superstitious man.

(Neve, 1892)

Politics of Health

Although a critical discussion on the politics of health during the 19th-century Kashmir is not the objective of the chapter, a side reference in this regard will not be out of context. Having discussed the scenario with Mark Harrison (Personal Communication, 1999), I am firm in my opinion that there could be a political motives for portraying Srinagar as filthy, for example. It is a fact that public health was seen as part of the British 'civilising mission', and that it was intimately connected to Imperial ideology pertaining to maximum resource exploitation. Imperial policy of cholera control was not based on a humanistic perspective, but as a result of economic reasons. To prove this point, I quote Lawrence:

I have stated above that if the disease is once established in the congenial filth of Srinagar it soon spreads to the villages, causing heavy mortality among the revenue-paying cultivators of the state, and just as I urge that, even in the interests of the land revenue alone, it is politic to prevent famine, and utter disorganization and financial ruin which attend on famines, so do I urge that it is financially wise to clean Srinagar, and to remove the present danger of cholera.

(Lawrence, 1895, p. 219)

This is a clear evidence that the efforts made by the government to control/eradicate cholera was aimed at to minimize loss of land revenue, other than welfare perspective.

Paradoxically, severe outbreak of cholera occurred even in the Military Cantonment of Lahore, which is considered cleaner than civil areas. According to Bellew:

out of the force comprising 2,452 men, women, and children, 880 were attacked with cholera, and 535 died in the space of little more than a month.....cholera hardly existed outside cantonments of the troops, and that, while our soldiers were dying by hundreds, the city close by, with its 90,000 people, remained almost entirely free from the disease.

(Bellew, 1885, p. 1)

The above examples clearly reveal that political and economic motives were behind the formulation and adoption of sanitary policies.

Conclusion

It is evident from the discussion that though insanitary conditions prevailed in Kashmir throughout the last century, the outbreak of cholera was confined to certain periods of time. The duration of cholera prevalence also varied from one month in 1827 to 13 months during 1875–76. Mortality figures due to cholera also varied from couple of thousands in 1867 to 18,000 in the 1892 outbreak. The study clearly shows that cholera was not indigenous to Kashmir, but was part of pandemic wave with a focus in the Bengal delta, sweeping across the Indian subcontinent and diffusing to Central Asian countries, Europe, Africa and the United States. The Study also argued that cholera diffused to Kashmir both via Punjab, i.e. Jhelum river route, as well as via Central Asian route including via Kabul. The study reveals that after 1892 cholera outbreak, strict sanitary measures and vaccination campaign started, resulting in the decline of a cholera incidence in Kashmir. Similar sanitary measures and vaccination campaigns were also launched in other parts of Indian sub-continent leading to the decline of cholera cases. Finally the study does highlight the significance of meteorological conditions in the outbreak of cholera in 1892, i.e. the drier weather during April to May followed by heavy rain and high temperature during summer months could possibly be the cause of cholera outbreaks. There are also indications of the relationship between the designation of Srinagar as filthy city and introduction of cholera control measures to enable the fulfilment of Imperial policies of resource exploitation.

Acknowledgement

I am grateful to The Wellcome Trust for providing Travel Grant for consulting research material in Libraries in London. I am also thankful to Dr. Mark Harrison of Director, Wellcome Institute of the History of Medicine, Oxford, for his critical comments. I am also thankful to the Reviewers of the chapter.

Notes

1 On the border of the plain valley of alluvium occur extensive elevated plateaus of alluvial or lacusterine material which occupy a great portion of the valley.

2 Lawrence's stress on the alluvial plains as a source of cholera may have been influenced by the 'sub-soil water' theory of the German Max von Pattenkofer, which was upheld by the British in India against Koch's bacilliary/water-borne (I.M.S. Officers ridicule Koch's findings, *Indian Medical Gazette*, 19, 1884, p. 14).

3 Dogra is the name given to the region around Jammu , and is said to be derived from a word meaning the 'two lakes', as the original home of the Dogra people was cradled between the lakes of Siroensar and Mansar. From Jammu stretching east along the plains of the Punjab the region is Dogra; and all who live in that tract are known as Dogras.

References

Bates, C.E. (1890), *A Gazetteer of Kahmir,* Reprinted in 1980 by Light & Life Publishers, New Delhi, p.463.

Bellew, H.W. (1885), *The History of Cholera in India From 1862 to 1881*, Trubner & Co. Ludgate Hill, London, p.502.

Clemow, F.G. (1903), *The Geography of Disease*, Cambridge Geographical Series, Cambridge University Press, Cambridge, p.93.

Drew, F. (1875), *The Jummoo and Kashmir Territories: A Geographical Account*, Edward Stanford, London, p.181.

Dutta, M.K. (1973), The diffusion and ecology of cholera in India, *Geographical Review of India*, Vol.35, pp.243–262.

Elmslie, W.J. (1875), *Seedtime in Kashmir: A Memoir of William Jackson Elmslie:* Thomson, W Burns. Publisher: James Nisbet, London.

Fayrer, J.C. (1888), *The Natural History and Epidemiology of Cholera*, London, pp.52–53; Also see, J. C. Fayrer (1894), Preservation of Health in India, London.

Howe, G.M. (1972), *Man, Environment and Disease in Britain*, Penguin Books, London, pp.178–179.

Hunter, J.M. (1988), Geography and Public Health, In *Earth'88: Changing Geographic Perspective.* Proceedings of the Centennial Symposium, National Geographic Society, Washington, DC, pp.117–118.

J&K Government. (1892), *Annual Administration Report*, Govt. of Jammu & Kashmir, Srinagar.

Jeffery, J.R. (1988), *The Politics of Health in India*, University of California Press, Berkeley, p.40.

Lawrence, W.R. (1895), *The Valley of Kashmir*, Henry Frowde, London, p.36.

Macnamara, F.N. (1880), *Climate and Medical Topography in Their Relation to the Disease Distribution of the Himalayan and Sub-Himalayan Districts of British India*, London, Longmans Green and Co.

Mitra, A. (1889), *Medical and Surgical Practices in Kashmir* (1889), Government Press, Lahore, p.9.

National Archives, Delhi, Kabul residents who are in the pay and employment of Cashmere, *Report*, Srinagar, *August, 1875*, pp.82–86 (National Archives, Delhi). Also see: Arrival of certain Afghan refugees at Srinagar, August, 1882, 1885, secret file (National Archives, Delhi).

Neve, E.F. (1892), Letter from Kashmir, *Extracts from Annual Letters*, November 9, Archives University of Birmingham.

Personal Communication. (1999), with Mark Harrison, Director, Wellcome Unit for the History of Medicine, University of Oxford, Oxford.

Pyle, G.F. (1979), *Applied Medical Geography*, V.H. Winston Sons, Washington, DC, p.129.

Raza, M. Ahmad, A. and Mohammad, A. (1978), *Valley of Kashmir*, Vikas, New Delhi, pp.82–85 (p.116).

Report from Dy. Commissioner on Special Duty at Cashmere to Govt. of Punjab, dated 10th December, 1861.

Report on the Sanitary Administration of the Punjab (1869), Government Press, Lahore, 1876, p.112.

Report from Dy. Commissioner on Special Duty as Cashmere to Govt. of Punjab, dated 10th December, 1999.

Tyndale Biscoe, C.E. (1892), Letter from Kashmir, *Extract from the Annual Letters*, October, Archives, University of Birmingham.

Tyndale Biscoe, C.E. (1921), *Thomas Russell Wade: A Pioneer in Kashmir*, Church Missionary Society, London, p.7.

Wilson, A.C. (1900), *Missionary to Kashmir: Irene Petrie*, Reprinted by Swati Publication, Delhi, 1993, pp.112–113.

3 Dr Henry Cayley's Pioneering Medical Work in Ladakh[1]

John Bray

Introduction

In late 1866 Sir Donald McLeod, the Lieutenant Governor of Punjab, put forward a proposal to send a British officer on a sensitive mission to Ladakh (House of Commons 1868, p. 5–6).[2] Sir John Lawrence, the British Viceroy, duly approved McLeod's plan, but noted that it might be difficult to find "an officer possessed of sufficient intelligence and discretion".[3] The main qualification would be "an aptitude for commercial and political enquiries". However, the best possible selection would be a medical officer who, in addition to other qualities, would be willing to "exercise his profession gratuitously among the people of Ladakh".

The man eventually selected for the post was Dr. Henry Cayley (1834–1904), who had studied at King's College Hospital in London, and joined the Indian Medical Service in 1857. Since then, after a stint as a military surgeon during the Indian Mutiny/Rebellion, he had served in civil posts in Gorakhpur, Simla and Burdwan.

Cayley opened a dispensary as soon as he arrived in Leh, the capital of Ladakh, in late June 1867. Over the following four and a half months, he treated a total of 659 patients. He therefore has the distinction of being one of the first Western-trained physicians to practise in Ladakh although not—as will be seen—the very first. This essay places Cayley in his historical context, showing how the themes of medicine and diplomacy complemented each other in early British engagement with Ladakh and Central Asia.

A Distinguished Predecessor: William Moorcroft in Ladakh

Cayley's posting marked a turning in point in British relations with Ladakh, but it was not entirely without precedent. The first British official to travel to Ladakh was William Moorcroft (1767–1825) who spent two years there in 1820–22, together with his younger companion George Trebeck (1800–25). Moorcroft was a veterinarian, in charge of the East India Company's stud: the official reason for his journey was that he hoped to travel from Ladakh to Yarkand and from there to Bokhara in search of Central Asian horses to replenish the Company's bloodstock. However, Moorcroft was also a physician, with a particular skill in cataract operations. For example, when passing through Hoshiapur in Punjab en route to

DOI: 10.4324/9781003329459-3

Ladakh, he operated on as many as 40 cases of cataract "with very fair success" (Moorcroft & Trebeck 1841, vol 1: p. 86).

Moorcroft's medical skills helped him to break down cultural and social barriers throughout his travels. Indeed, Christopher Bayly (1996, p. 136) argues that the "discourse of medicine" runs like a thread throughout his travels, letters and journals: "It was not simply that Moorcroft and his Indian informants constantly discussed medicine, but that their view of the world was fundamentally formed by it."

After reaching Leh in September 1920, Moorcroft lost no time in opening his medical practice and soon had more patients than he could cope with (Alder 1985, p. 259). Among other measures, he tried to introduce vaccination against smallpox, but was unsuccessful because he was unable to obtain a sufficient quantity of active vaccine from his friends in India. Overall, however, his medical practice proved markedly successful and, as Alder (1985, p. 260) points out, his doctoring "gave him access to some of the most important men in the Ladakhi political and religious establishment".

Moorcroft's skill as a healer was no doubt an important factor in helping him win the trust of the Ladakhi Kalon (*bka' blon*—minister) with whom he drafted a letter seeking to place Ladakh under British protection (Moorcroft & Trebeck 1841, p. 418–422). However, at this stage the East India Company had no interest in extending its political reach so far to the north. To Moorcroft's dismay, the British authorities in Calcutta dismissed his proposal and he was "severely censured for taking unauthorisedly a part in political arrangements" (ibid., p. 421). In late 1822 he moved on from Ladakh to Kashmir and from there via Afghanistan to Bokhara. He died in 1825 on his return journey.

Growing British Interest in Central Asia

The Kalon's motive in seeking British protection was to safeguard Ladakh against the expansionist ambitions of Ranjit Singh (1780–1839), the Maharaja of the Sikh Empire. This proved to be a well-founded concern. In 1834 Raja Gulab Singh of Jammu (1792–1857), a feudatory of the Sikh empire, sent an army to conquer Ladakh. In 1842, after eight years of conflict, Ladakh finally and decisively lost its independence to Jammu. In 1846, the British rewarded Gulab Singh for his support in the First Sikh War by installing him as the first Maharaja of the combined princely state of Jammu & Kashmir, including Ladakh.

Initially, the British hoped that Jammu & Kashmir would serve as a buffer state on the northern frontiers of India, and at first they took no great interest in its internal administration or in its links with Central Asia. This began to change in the course of the 1860s. One factor was a greater enthusiasm on the part of the Punjab Government for the prospects for trade with Central Asia as expressed in an official report by R.H. Davies (1862). A second concern was the need to monitor political developments to the north. Already in the 1820s Moorcroft had pointed to the potential threat to British interests from Russian expansion in Central Asia.

Cayley was therefore assigned three main tasks for his mission to Ladakh.[4] The first was to enquire into the tariffs imposed by the Jammu & Kashmir government on trade passing through Ladakh. Second, he was to enquire closely into trade opportunities between India and Central Asia. Third, he was to "pick up and sift all the political information that might come his way", particularly as regards developments in Eastern Turkestan.

It would be good to think that the Government of India had Moorcroft's travels in mind when they proposed that a medical officer would be best suited to carry out these tasks. However, it is more likely that they were thinking of more recent precedents. Cayley's obituary in the *British Medical Journal* (Anon 1904) suggests that his appointment was a "tribute to the powers and conciliation of the Indian Medical Service" and that this was "justified by repeated experience of the humanizing influence of medicine and the popularity of medical men on the Punjab frontier." This appears to be a reference not to Ladakh but to India's north-west frontier with Afghanistan.

Cayley's Medical Practice

In his official report to the Punjab government, Cayley explained that he reached Leh, the capital of Ladakh, on 24 June 1867, and at once made it known that all applicants would receive medical treatment.[5] He had brought with him a small supply of medicines and instruments, and was accompanied by a hospital compounder named Khuda Buksh who served as a "native doctor". He converted two small tents into a hospital and reports that a "grove of poplar trees served as an operating theatre", adding that "for surgical assistants numerous Ladaki amateurs were always at hand, who took great interest in the proceedings".

During the first few days, patients came freely, but then the attendance suddenly ceased almost completely apart from a few sepoys (soldiers from the local Kashmir garrison) or other government servants. Kashmir officials in Leh had accepted Cayley's presence in Ladakh only reluctantly, and wished to discourage local people from coming into contact with him. They even went so far as to sponsor a rival doctor, a Kashmiri *hakim* (practitioner of the Unani Greco-Islamic system of medicine). For a time, patients on their way to Cayley's dispensary were "forcibly stopped" and taken to his competitor. Cayley protested vigorously, and the flow of patients resumed.

Even then, Cayley still had to negotiate significant social obstacles. These included the fact that people were unfamiliar with Western medicine and therefore had to "discover for themselves that they derived any benefit by attending". Moreover, Ladakhis were extremely poor and with therefore found it difficult to leave their work however ill they might be. A further obstacle was that many of those who did attend were carried off to perform enforced labour.

Against this background, Cayley felt some satisfaction in the relatively large number of patients who attended his dispensary, as shown in these statistics (Table 3.1):

Table 3.1 Number of patients who attended his dispensary, as shown in these statistics

	25–30 June	*July*	*August*	*September*	*October–6 Nov*	*Total*
Admissions	22	239	140	120	129	659
Daily average	6	30	29	26	22	27.5

Table 3.2 Geographical backgrounds of admissions

Ladakhis	260
Sepoys	97
Baltis	84
Yarkandis/Turkis	71
Argons (mixed race)	49
Kashmiris	45
"Hindustanis"	35
Men from Kulu and Lahoul	14
Men from Lhasa	5
Kabulis	3
Others	5

He states that 532 patients were male, and 126 female, while there were only ten children under 12 years of age. They came from a wide range of geographical backgrounds (Table 3.2).

Of the total number, 138 were traders and 37 pilgrims from Central Asia—here referring to Eastern Turkestan rather than Tibet. It seems that one man came all the way from Yarkand on the far side of the Karakoram mountains seeking treatment for a chronic illness. By contrast, the low number of patients from Lhasa reflects the fact that there was a much larger volume of long-distance trade with Turkestan than with Central Tibet, especially in the summer months.

Pattern of Diseases

Cayley likewise kept a statistical record of the diseases that he treated. These figures cannot be taken as a comprehensive medical survey because of the manner in which they were collected: all the patients were volunteers and, despite his success in out-manoeuvring the Kashmiri *hakim*, other would-be patients may have preferred to stay away, or to seek treatment from indigenous practitioners.[6] Nevertheless, his figures present a striking "snapshot" of medical conditions at the time (Table 3.3).

Cayley notes that dyspepsia "of a most obstinate and troublesome nature" ranks as one of the chief diseases of the land, and suggested that it was caused by bad diet: the "everlasting and unvarying *suttoo*"—roasted barley flour, better known in the wider Tibetan world as *tsampa*. He added:

One sees a man with a lump of uncooked dough as big as his head, and this he swallows in large pellets, washing them down with cold water, and this constitutes his sole diet for days together.

In further comments on the Ladakhis' diet, Cayley noted that there were nine cases of scurvy—a disease caused by lack of vitamin C—and that many other patients had shown signs of the same deficiency:

During the early part of the summer the majority of the sick showed some sponginess of the gums, owing, doubtless to the absence of fresh vegetables and other anti-scorbutic elements in their diet. In the upper parts of Ladak

Table 3.3 Pattern of diseases

Dyspepsia	99	Dysentery	7
Fevers	92	Abscess	7
Venereal diseases	69	Entispion	6
Ophthalmia	64	Epilepsy	5
Rheumatism	57	Diarrhoea	3
Caries of teeth	39	Staphyloma	3
Bronchitis	28	Scabies	3
Ulcer	20	Cephuloea	3
Accidents	19	Amaurosis	2
Colic	16	Piles	2
Cataract	14	Frost bite	2
Tumours	11	Ascarides	1
Deafness	11	Hepatitis	1
Skin diseases	10	Erysipelas	1
Scurvy	9	Other diseases	17
Paralysis	8		

and about Leh, where fruit is not plentiful, the food from October to June consists almost entirely of *suttoo* and water, and a few dried herbs, and I certainly saw more cases of spongy gums in June and July than later in the year, when herbs and common vegetables were plentiful.

At first sight "fevers" constituted a large proportion of the sicknesses that Cayley treated but he explained that:

> … the 65 cases were nearly all of a most trifling nature, the attack only lasting two or three days, and generally depending on indigestion, cold, exposure to the sun whilst at work, standing in cold water, and such like causes.

Venereal diseases were very prevalent, notably among the sepoys, but Cayley noted that the symptoms seemed to be less severe than in the plains of Indian or in Europe, and suggested that this might be something to do with the dryness and antiseptic properties of the air.

Opthalmia—inflammation of the eye—was very common:

> The disease appeared in many cases to have risen from the glare of the snow whilst crossing mountain passes (as a protection from which the natives often wear snow spectacles made of plaited hair); in others, from exposure to the intense glare and dust and heat of the sun, in the barren sandy deserts which extend over so much of Ladak.

In perhaps unconscious emulation of Moorcroft, Cayley carried out 14 cataract operations and reports that this made a great impression:

> …on my return journey the blind were brought from all sides to meet me on the road, though, of course, I could not then perform such an operation.

Cayley does not seem have treated any cases of smallpox, but notes that it had been a severe problem in the recent past:

> Ten years ago it spread through the whole country, and killed numbers: the whole population was inoculated in that year by the Lamas, and since then the disease has not occurred, excepting in a few cases last year.[7] In former years the custom was to expose the patients with the disease out on the mountain sides, where the friends brought them food, &c. until they either died or got well. It was a somewhat cruel, but, at the same time, admirable, plan for lessening the spread of the disease: and in this climate it would really be better for the sick to be out in the open air, than shut up in a close dwelling. Since the general inoculation ten years ago, the dread of the disease has greatly diminished.

Again echoing Moorcroft's earlier ambitions, he suggested that vaccination might be introduced without difficulty.

Finally, it is striking that diarrhoea and dysentery were almost unknown: the cases that came for treatment were nearly always pilgrims and traders from other regions rather than indigenous Ladakhis.

Information Gathering, Diplomacy and Trade

Cayley comes across as a doctor who was both conscientious and humane, but the main reason for his presence in Ladakh was not the practice of medicine, but rather a requirement to monitor trade and political developments in Ladakh and Eastern Turkestan. This was not a straightforward task. As the ruler of a princely state within the Indian empire, Maharaja Ranbir Singh (r. 1857–85) was obliged to recognise British paramountcy, especially in matters of foreign policy. Nevertheless, he still aspired to maintain his own lines of communication with the rulers of Central Asia, independently of the Government of India. He therefore responded to British pressure to appoint an officer on special duty in Ladakh only with the greatest reluctance, and Kashmiri officials in Ladakh did their best to impede Cayley's access to independent sources of information. For example, in October 1867 Cayley reported that:

> Obstacles are still secretly thrown in my way to prevent the traders freely communicating with me. The Yarkand merchants are on their arrival privately told that they had better not give information or make any complaints to me, as their doing so will be displeasing to the Wazir and to the Maharajah...[8]

In these circumstances, his medical activities provided him with a significant advantage because:

> ...the opportunity the dispensaries afforded me of communicating with the traders and others, and of conversing freely with them, were very great, and this I found of the greatest advantage in enabling me to pick up information.[9]

The British authorities evidently regarded Cayley's posting to Ladakh as a success because he was reappointed to serve in Leh for three more seasons. In 1870, the Government of India went a step further by negotiating a Commercial Treaty with the Maharaja of Kashmir (Drew 1875, p. 447–452). Under the terms of the treaty, the British would from now on appoint a Joint Commissioner every year: his task would be to spend the summer months in Ladakh to supervise the trade routes from Yarkand, in association with a Commissioner from the Kashmir government.

Epilogue

Cayley relinquished his Ladakh appointment in 1871 and took no further part in Ladakh or Central Asian affairs. Nevertheless, he occupies a special place in the history of Ladakh because he was the first of a series of British Joint Commissioners who continued to travel to Leh each summer until 1946, shortly before Indian independence.

Cayley's immediate successor was Robert Shaw (1839–79), a tea planter who had travelled to Yarkand the previous year and now entered full-time government service. Shaw served in Leh in 1871 and again in 1873. The one-year gap in 1872 was filled by J. E. T. Aitchison (1835–1898) who, like Cayley, was an officer of the Indian Medical Service.

To my knowledge, none of the subsequent Joint Commissioners were doctors. However, the British authorities continued to sponsor a medical dispensary in Leh. For example, in 1880 British Joint Commissioner Ney Elias reported that there had been a total of 562 treatments in 1877, 943 in 1878 and 938 in 1879.[10] One Chirag-ud-din served as Hospital Assistant while a lama named Sonam Tandup had made a practice of obtaining smallpox vaccine from the British dispensary. Travelling from place to place during a great part of the year, the lama had conducted 634 vaccinations in 1877, 347 in 1878 and 445 in 1879. The lama received no salary, and Elias suggested that he should be given Rs 15 for his work over the previous year. Moorcroft would have been delighted that, in this somewhat unexpected manner, his dream of establishing the practice of vaccination had at last been fulfilled.[11]

In late 1886 Karl Marx, a German doctor who had been trained in Edinburgh, joined the newly founded Moravian mission in Leh. His first operation was on a lady from Zangskar seeking treatment for cataract (Marx 1887, p. 162)—a continuation of what was now becoming an established tradition of Western eye surgery. The operation itself went well but at this stage Marx had no premises in which to offer residential aftercare and, since his patient was staying in a cold, smoke-ridden room, he feared for the outcome. Fortunately, she healed well, and in April 1887 Marx was able to take over the British dispensary with the help of a subsidy from the British government. He now had space to take on resident patients, who came from a similar range of geographical backgrounds to Cayley's first intake in 1867 (Marx 1890, p. 227). Marx himself died tragically young in 1891. However, he was the first of a series of Moravian medical practitioners to serve in Leh over a period of more than 60 years.[12]

As for Cayley himself, his first step after leaving Ladakh was to take a 13-month furlough during which—perhaps as a result of his experiences in Leh—he made a special study of ophthalmology at Moorfields Eye Hospital in London (Anon 1904, p. 812). Thereafter, he returned to Bengal and became Professor of Ophthalmic Surgery at Calcutta Medical College. After his return to Britain in the late 1880s, he served for seven years as Professor of Military and Tropical Medicine at the Army Medical School in Netley, near Southampton. Among other honours later in his life, he was appointed Honorary Surgeon to the Queen in 1891. Later, he volunteered to serve as a doctor in the 1899–1902 South African war, and died in retirement in 1904.

Cayley's own verdict on his first year of medical operations in 1867 was that they had been "on the whole successful, and have certainly been the means of relieving much suffering".[13] He suggested that many of his patients felt a "very lively gratitude", and added that "such a feeling is calculated to spread abroad a spirit of friendship towards India among the races beyond the Himalayas, and thus indirectly promote future trade".[14]

Cayley's observations point to wider continuities in British engagement with the Himalayan region, including Tibet as well as Ladakh. For example, the doctors who accompanied British missions to Tibet in the 1930s and 1940s provided free medical treatment to local people. As Alex McKay (2008, p. 123) points out, this practice "was intended to engender indigenous goodwill towards the British that could be translated into diplomatic advantage." In all these cases, humanitarian concerns were accompanied—indeed overlaid—by political and commercial motives. It would therefore be claiming too much to suggest that the relief of suffering was a primary goal in the official mind. It is nevertheless reassuring to note that it remained a constant theme.

Archival Sources

Cambridge University Library. Department of Manuscripts. Add. MS 7490. Mayo Papers.
Royal Geographical Society Archive, London. Ney Elias Collection.

Notes

1 An earlier, slightly longer version of this paper was published in *Tibetan and Himalayan Healing. An Anthology for Anthony Aris* (Edited by Charles Ramble and Ulrike Roesler. Kathmandu: Vajra Books) pp. 81–96. I am grateful for the editors' permission to republish.
2 T.H. Thornton, Secretary to Government of Punjab, to J.W.S. Wyllie, Lahore, 13th December 1866. Great Britain. House of Commons (hereafter cited as HoC) 1868, p. 3.
3 J.W.S. Wyllie to Secretary to the Government of Punjab, 22 January 1867. HoC 1868, p. 4–5.
4 J.W.S. Wyllie to the Secretary to the Government of Punjab, 22 January 1867 (HoC 1868, p. 4–5).
5 The main source for this section is: Cayley to the Government of the Punjab, 23 December 1867 (HoC 1868, p. 22–25). This is supplemented by further details drawn from his report in the *Indian Medical Gazette* (Cayley 1867–68).

6 For an earlier discussion of medical conditions on Ladakh referring to Cayley, see Akhtar (2007) who draws on the figures from June to September published in Cayley (1867–68). In this essay I cite the figures running up to November as published in HoC (1868).
7 This is the only case in the sources available to me where Cayley refers to doctoring by Buddhist rather than Muslim practitioners. The practice of inoculation using cowpox vaccine was pioneered by the British doctor Edward Jenner in the late 18th century, Lobsang Yongdan (2016) shows that the Fourth Tsenpo Nomönhen (Jampel Tendzin Trinlé) introduced this technique into the Tibetan medical tradition as early as the 1830s. It is not clear whether Cayley is here referring to Jennerian inoculation or to variolation, the practice whereby doctors immunize an individual against smallpox with material taken from a patient in the hope that a mild but protective infection would result. This technique had been known to the Chinese since the 15th century.
8 Extract from Dr Cayley's Diary, dated 1st October 1867. HoC (1868, p. 16).
9 Cayley to the Officiating Secretary of the Government of the Punjab, 23 December 1867. HoC (1868, p. 25).
10 Trade and Hospital Statistics of Ladakh, dated Leh 6th October 1880. Included in Ney Elias papers, Royal Geographical Society Archive, London.
11 For a further discussion of vaccination, see McKay (2008, pp. 134–142).
12 For an overview of the Moravian mission in Ladakh, see Bray (1985). For missionary doctoring and a discussion of the role of the Indian Medical Service in 20th century Tibet, see McKay (2008).
13 Cayley to Officiating Secretary to the Government of Punjab, 23 December 1867. HoC (1868, p. 24).
14 Ibid.

References

Akhtar, R. (2007). Changing disease ecology of Leh District: Contemporary scenario and historical perspective. *Punjab Geographer* 3, pp. 39–44.
Alder, G. J. (1963). *Beyond Bokhara. The Life of William Moorcroft*. London: Century Publishing.
Anon. (1904). Deputy-Surgeon-General H. Cayley, F.R.C.S., C.M.G., Honorary Surgeon to the King. *British Medical Journal*, 1(2257), pp. 811–812.
Bayly, C. A. (1996). *Empire & Information: Intelligence Gathering and Social Communication in India, 1780–1870*. Cambridge: Cambridge University Press.
Bray, J. (1985). A history of the Moravian Church in India. In *The Himalayan Mission*, pp. 28–75. Leh: Moravian Church.
Cayley, H. (1867–1868). Notes on Ladakh in 1867. *Indian Medical Gazette* 2, pp. 266–268; 3, pp. 3–5.
Davies, R. H. (1862). *Report on Trade and Resources of the Countries on the North-Western Boundary of British India*. Lahore: Government Press.
Great Britain. House of Commons. (1868). East India (Ladakh). Copy of Correspondence between the Secretary of State for India and the Governor General in Council, relating to the Appointment of a Commercial Agent in Ladakh, and to his Proceedings there. Ordered, by the House of Commons to be printed, 16 March 1868. Command No. 147. London. HoC: see Great Britain. House of Commons.
Lobsang Yongdan. (2016). The Introduction of Edward Jenner's Smallpox Vaccination into Tibet. *Archiv Orientální* 84, pp. 577–893.
Marx, K. (1887). Westhimalaya. *Missions-Blatt aus der Brüdergemeine* 51, pp. 161–167.
——— (1890). Britisch-Indien. Für Leib und Seele! *Missions-Blatt aus der Brüdergemeine* 54, pp. 224–231.

McKay, A. (2008). *Their Footprints Remain: Biomedical Beginnings Across the Indo-Tibetan Frontier*. Amsterdam: Amsterdam University Press.

Moorcroft, W. and Trebeck, G. (1841). *Travels in the Himalayan Provinces of Hindustan and the Panjab, in Ladakh and Kashmir, in Peshawar, Kabul, Kunduz and Bokhara*. Ed. H.H. Wilson. 2 vols. London: John Murray.

4   Arthur Neve (1859–1919) and a Mission Hospital in Srinagar, Kashmir

Rais Akhtar

Introduction

Kashmir had undergone various phases of historical and political development since the beginning of the 19th century, and therefore socio-cultural economic and political conditions resulted in wide variation in the attitudes towards health- and medical-care systems in Kashmir.

The Arabs and the early Muslims attached great importance to the care of the body, thus cleanliness and health were acts of healing and piety. Sultan Zainul Abedin, the 'Akbar of Kashmir (1423–72 AD), ordered the translation of several medical works and patronized among others the great physicians Sribhatta and Mansur Bin Ahmad' (Askari, 1957). The Unani system of medicine together with spiritual medicine was the dominant system in the valley of Kashmir until 1865 when the missionary doctor, Dr William Jackson Elmslie (1832–72) established an allopathic dispensary in Srinagar.

Although some modern allopathic physicians had visited Kashmir during the early 17th century and especially Dr F Bernier (1625–88), a French physician who was a personal physician to the Mughal Emperor Aurangzeb (1618–1707), Dr Elmslie was the first medical missionary who arrived in Kashmir in 1864.

Subsequently other medical missionaries, including the brothers Dr Arthur Neve and Dr Ernest Neve, spent several years in Kashmir during the later half of the 19th and the early part of the 20th centuries. The diffusion of modern medicine through medical missionaries led to a conflict with the local Hakims – practitioners of Unani medicine – who protested before the Maharaja of Kashmir not to grant favour to medical missionarie.

Dr Elmslie was the first medical missionary who arrived in Kashmir in 1864. Subsequently other medical missionaries, including the brothers Dr Arthur Neve and Dr Ernest Neve, spent several years in Kashmir during the later half of the 19th and the early part of the 20th centuries. The diffusion of modern medicine through medical missionaries led to a conflict with the local Hakims – practitioners of Unani medicine – who protested before the Maharaja of Kashmir not to grant favour to medical missionaries.

DOI: 10.4324/9781003329459-4

Socioeconomic Conditions

Poor economic conditions and insanitary and unhygienic living conditions present an environment conducive to disease. Various infectious diseases and Kangri burn cancer (specific to the region) were widely prevalent in Kashmir. The report dated 10 December 1861 of the Deputy Commissioner on Special Duty at Kashmir to the Secretary to the Government of Punjab bears testimony to the poor economic conditions of the people in Kashmir. The report states: 'The people in Kashmir are wretchedly poor and in any other country their state would be almost one of starvation and famines' (Report, 1861). Under such socioeconomic circumstances it was not possible for people to maintain a hygienic environment. Dr William Jackson Elmslie described in his letters the unhygienic practices of the people that result in the widespread incidence of skin diseases. In one letter he writes:

But there are other things in Kashmir which most terribly detract from its pleasure as a place of residence. The dirt is beyond description. Who can tell what Kashmir smells are? Not the odour of roses, such as one has expected to fill the air; but oh! Such, that the dirtiest of London Courts is sweeter than the cleanest of Kashmir villages. The clothes, too, of the people are filthy; not that the filth shows much, for all their garments are of grey wool, which is a most perfect concealer of dirt; but not a few of their diseases are the result of their uncleanliness and how often I have almost shrunk away from them as, in my dispensary, while I have been examining a patient, I have seen the lice crawling on his clothes and his fleas skipping over to me. Of course, if you can avoid all intercourse with the native, than dirt is not such a continual source of annoyance, but to us it was a daily trouble. The heat evolved is often considerable. According to Elmslie the injurious effects of the heat of the Kangri cause the cancers of the abdomen and thighs.

Medical Work (Research)

One paper published in The Lancet in 1891 by the Neve brothers described the number of surgical operations in Walter Lawrence (1857–1940), a Settlement Commissioner in Kashmir, also noted in 1895: 'The clothes of the villagers are simple and extremely mean in appearance'.

The Hospital

The hospital was constructed basically under duress from the British Resident in Srinagar, to allow the Medical Mission in Kashmir. It was simply a wooden shed and not even rainproof. In 1864 Robert Clark (1825–1900) of the senior Christian Missionary Society in the Punjab with his wife crossed the mountain passes into the far-famed valley of Kashmir during the hot season and with characteristic energy and without seeking anybody's help in 1865 quietly opened a dispensary to be run in Srinagar by his wife on her own. She was not a qualified doctor but she seemed to know more than the native Hakims and soon a hundred patients were attending each day. Poverty-stricken Kashmir was the home of communicable diseases and

Figure 4.1 Kashmir Mission Hospital Established by Arthur Neve

unhygienic conditions, and skin diseases were widely prevalent. Here Dr Elmslie worked throughout his life in Kashmir: Yes and this was Elmslie's first building, grudgingly erected by the State. Another shade was promised, but it had not materialized. This, then, was the first Mission Hospital and the only building in Kashmir deliberately constructed with a view to extending the benefit of western medical skill and science to the people of Kashmir (Neve, 1928, Figure 1).

Kangri Burn Cancer

Dr Elmslie conducted research on Kangri burn cancer and published the first ever paper on the subject in the Indian Medical Gazette and his book, *Seed Time in Kashmir: A Memoir* (Elmslie, 1866, 1875). Cultural tradition and climatic necessities work together in the continuance of a centuries-old practice of carrying Kangri, a portable heat source, under the clothes from about the beginning of November up until late March. The Kangri, an earthenware bowl five or six inches in diameter, is surrounded by willow basket work and surmounted by a wicker handle. It contains small quantities of lighted charcoal, sawdust and cleaner leaves. Elmslie noted 'when the weather is extremely cold it is customary for both men and women while walking about out of doors to carry the Kangri under their loose woollen gowns and in close proximity to the bare of skin of the abdomen'. When indoors or in a sitting posture Kashmiris place the Kangri between their thighs (Akhtar, 1992). The fuel consumed in the Kangri is charcoal and the heat evolved is often considerable. One paper published in The Lancet in 1891 by the Neve brothers described the number of surgical operations in the Mission Hospital during the 1882–89 period (Neve and Neve, 1891).There were 3651 eye operations, 864 operations for tumour and 579 bone operations. Details of the population and health conditions in Kashmir, well-to-do peasants and somewhat less thriving townspeople, were described. The climate was temperate with cold snowy winters and warm summers. The chief diseases were syphilis and smallpox.

In 1923 Ernest Neve wrote another paper on 'Kangri burn cancer' that was published in the British Medical Journal. Since 1881, operations for epithelioma performed in the Kashmir Mission Hospital (Figure 4.1) numbered 2491 of which approximately 2000 (84%) were for Kangri cancer. The disease was more common in men than in women, probably owing to the less frequent use of the Kangri by the latter in their mostly domestic occupations. Ernest Neve asserts that while operating on cases of Kangri burn cancer the glands should be dealt with in the first place because the primary tumour is always septic. Recurrence occurred in some 20% of the operated patients (Neve, 1923).

Arthur Neve, Physician and Traveller

Arthur Neve wrote several accounts of his journeys in the Kashmir and Central Asian regions. Most of his geographical accounts were published in the Royal Geographical Society's Geographical Journal. For these contributions Neve was awarded the Beck Prize in 1911. In his book Thirty Years in Kashmir published in 1913, Arthur Neve provides us with an account of all his explorations and mountaineering expeditions in Kashmir (Neve, 1913). He travelled far and wide in that part of the British Empire 'where three empires meet' and all who are interested in Kashmir and the great mountain ranges that lie to the north of it will find in this book plenty of information, adventure and interesting reading (Geographical Journal, 1911, 1913).

Sir Walter Lawrence (1857–1940) writes of Arthur Neve's research findings:

I remember a paper written by Dr Neve and published a few months ago. He proved the existence of three parallel ranges in the far west of the system of the Karakoram. Beyond doubt two of them belong to the most gigantic systems on the Earth's Surface.

(Geographical Journal, 1911)

Some of the country he describes is the most magnificent in the world. The great range of Karakoram Mountains, the unique Indus Valley, the lofty Himalayan peak Nanga Parbat and the twin peaks of Nun Kun are among the most important of his subjects. He explored new country, crossed passes not visited Mission Hospital, Srinagar, 1882. Dr Arthur Neve is in the centre surrounded by patients and hospital staff previously by Europeans, climbed several times to over 20,000 feet and was always ready to help the natives with whom he came into contact, offering medical advice and performing small operations such as were possible in the circumstances (Arthur, 1913).

Sir Walter Lawrence, who had spent six years in Kashmir and published his famous book – The Valley of Kashmir in 1895, further commented on the work of Arthur Neve:

> If I had had the privilege of listening to Dr Neve before [I] attempted to write my poor chapter on Physical History in The Valley of Kashmir, I should have been able to write with much greater effect. From what I have seen myself, I can testify to the great accuracy of Dr Neve's account. I think this is a good opportunity, before this great audience, for letting you know what the work of the Medical Mission in Kashmir is doing. I lived six years in that country and know the road from Kashmir to Gilgit and Ladak. Wherever I went there was only one question. The people did not want to see me but they wanted to know when Neve Sahib was coming – Neve Sahib who brought comfort and healing wherever he went. Working with very little help, working in a very small way against every hindrance, against the Brahman influence, the two Neves (Arthur Neve and Ernest Neve) have won everything to them and now they have a grand hospital in Srinagar and when the Neves are not going into the villages, the villagers are coming in to the Neves.
>
> (Geographical Journal, 1911)

Ernest Neve described his commitment of Arthur Neve to medical work:

> Dr Arthur Neve took charge of the hospital in 1881. In that year 10,800 new patients were treated; there were 23,393 patient visits, and 1418 operations were performed. Since then the Medical work has progressed. The original mudbuildings have gradually been replaced by solid masonry structures. And the steady growth of the number of in-patients, and the readiness with which even upper-class women remain in the hospital, testify to the confidence with which the institution is now regarded. It is now renowned through all the north India and is a splendid the testimony to the steadily, thorough and persevering work of two self-sacrificing men
>
> (Neve, 1928)

Writing about the skill of Dr Arthur Neve, Dr Henry Holland (1875–1965) wrote in his autobiography (Holland,1958):

> so great was the rush and so large the number of patients who flocked to us (in Shikarpur near Sukkar) that I sent an SOS to Dr Arthur Neve who was then on a tour in the Punjab. A distinguished surgeon, he came to the rescue and on a single day (5 December 1909) per- formed forty-seven eye operations; without his help the work could not have been undertaken.

Arthur Neve wrote:

> The mission hospital has been completely rebuilt and considerably enlarged
> to accom- modate over 100 beds ... We have also erected and superintend the
> large Leper Asylum. During the past ten years over 300,000 visits have been
> paid to the hos- pital by patients, and 30,000 surgical operations have been per-
> formed. These figures prove that the people of Kashmir, both Mohammedan
> and Hindu, appreciate the work that is being done whether by clergy or doctors
>
> (Neve, 1899)

Writing on 4 June 1892, during the severe outbreaks of cholera in Kashmir, Neve
described the horrific situation and the panic of cholera gripping Srinagar City. He
writes for the Medical Mission at Home and Abroad:

> But during the following week the increase was very great, and there was
> rather a panic. The Resident went off to Gulmarg, and everybody followed
> as fast as coolies and ponies could be obtained. Here we quietly settled down
> to fight the epidemic with means available. As far as the European quarters
> were concerned, this was comparatively easy; but the habits and prejudices
> of the people fostering the spread have suffered frightfully. The deaths rap-
> idly increased from five or 10 a day to a 100, then to 200, then to 300 a day,
> and a large proportion of those attacked were buried the same day
>
> (Neve, 1892).

In 1919, the British Medical Journal published an Obituary: 'when off duty, in
his few holidays, his favourite relaxation was mountaineering in the Himalayas'.
(Obituary,1919) Arthur Neve writes in his book Picturesque Kashmir that even during holidays people come to know about Neve's visit even in difficult inaccessible
areas of the Himalayas and requested treatment and operations:

> Patients followed even to the Shayok, and so fervent were their petitions that
> I con- sented to operate for cataract while waiting for our baggage to be fer-
> ried over the river. In order to sterilize my instruments by boiling, fire had to
> be obtained, in the absence of matches, by recourse to flint and steel, a little
> gunpowder and a bit of rag torn on the spot from a man's shirt. A primitive,
> but satisfactory expedient!
>
> (Neve, 1900)

Hospital Routine Work (1877–83)

From 1877 to 1883 Kashmir was severely affected by famine and cholera. The work
at the hospital continued uninterrupted. Dr Edmund Downes (1842–1911) wrote:

> The famine, and the horrors connected with it, and the general misrule in
> Kashmir, were obstacles to the advance- ment of any good cause ... Many

patients were brought in too far gone for any hope, and the starving people often made poor recoveries ...

I found it most convenient to see the patients about midday, leaving the morning free for assistant surgeon and dressers to do their work while I could attend to any special patients or other work. At about 12 O'clock I went to the hospital and at once took my place on the veranda in front. By this time a good number would be already collected and seated under the shade of a rough awning, built over the ground just in front. Others could be seen coming from the three roads that converged at the foot of the hill on the side of which the hospital stood. Some who had come by boat through the Dhal Lake found their way through the village of Drogjan and ascended the little hill to the hospital; any who were very ill would be carried on the rough light beds of the country; others would come down the poplar avenue walking, or riding ponies, or sometimes carried in beds; others would arrive similarly from the country villages by the road on the north side. Blind people came led by children, women with babies or little children in their arms, despairing in their grief, loud in their emotion, struggled painfully onwards to try the last resource of the English physician (CMS, 1877–83).

This account shows the lack of effectiveness of the traditional system of medicine in Kashmir as a result of which patients are left without any option but to go to hospital to be treated by English physicians.

The Impact of the Mission Hospital

In 1866 when Dr Elmslie started operating at the Mission Hospital, there was a dispensary run by the Maharaja of Kashmir and based on the indigenous systems of medicine – Unani as well as Ayurveda. Encouraged by the western (allopathic) medicine of missionary doctors at the Mission Hospital in Srinagar and its popularity among the Kashmiris, the Government's (Maharaja's) allopathic dispensaries were established in different districts in Kashmir. Maharaja Ranbir Singh (1830–85) who ruled Kashmir from 1858 to 1885 established a regular medical department and in the closing years of his reign set up 27 medical institutions. He encouraged both Unani and Ayurvedic systems of medicine in the state. Later the Maharaja also introduced the allopathic system and appointed Dr Bakshi Ram Head of the Medical Department. Adopting the missionary approach, the patients of the Maharaja's dispensaries provided free treatment and inpatients were given free food, clothes and bedding according to the season of the year. Both Hindu and Muslim servants were required in the hospitals to attend to the patients of the two communities undergoing treatment (J&K Govt.1882-84).

Conclusions

In many countries hospitals were established within colonial motives but in Kashmir first a dispensary and later a Mission Hospital were established by the Medical Missionary Society in order to provide modern medical care to the population without much disturbing the availability of the indigenous system of medicine.

Despite the practice of the Unani system of medicine, allopathic medicine when first introduced into Kashmir became extremely popular among the people. The missionary doctors who worked in the dispensary and hospital, and particularly Arthur Neve, were committed to serving the population and were devoted researchers who published their results in internationally reputable journals – The Lancet, the British Medical Journal and the Indian Medical Gazette. Arthur Neve also wrote Kashmir, Ladakh and Tibet (1899), Picturesque Kashmir (1900) and Thirty Years in Kashmir (1913).

Arthur Neve died suddenly in Srinagar of a fever on 5 September 1919. The Maharaja of Kashmir ordered a State Funeral. The mourners included people from all classes, races and religions who united to pay tribute to their well-beloved and disinterested friend who had done so much for Kashmir and its people. The Obituary in the Geographical Journal in 1919 noted: 'He probably did more than anyone who ever lived towards the amelioration of suffering for various native races of that country among whom his reputation was extraordinary' (Longstaff, 1919).

Acknowledgements

The author is grateful to the Wellcome Trust for a Grant and to the Royal Society for a Short Visiting Fellowship that enabled him to visit London, Birmingham and Oxford libraries. He also thanks Professor Mark Harrison of the Wellcome Unit of the History of Medicine, University of Oxford, for his useful suggestions.

References

Akhtar R. (1992) Kangri. Traditional personalized central heating in the Valley of Kashmir. *Focus*;42:2127.

Askari H. (1957) Medicines and hospitals in Muslim India. *Proceedings of the Indian History Congress. Medieval India*:174.

Geographical Journal. (1911) Journeys in the Himalayas. *Geographical Journal*;38:357.

Geographical Journal. (1913) Thirty years in Kashmir, Review of the book by Arthur Neve. *Geographical Journal*;43:429.

CMS. (1877–83) The opening of the door in Kasmir-1877 – 83. Christian Missionary Society, Medical Missions Report. Literary Archives, University of Birmingham:252–53.

Elmslie WJ. (1866) Etiology of epithelioma among Kashmiris. *Indian Medical Gazette*;I:324–326.

Elmslie WJ. (1875) *Seedtime in Kashmir: A Memoir*. London: James Nisbet and Co.

Holland H. (1958). *Frontier Doctor: An Autobiography*. London: Hodder and Stoughton: p. 118.

J & K Government. (1882–84). Biennial Report of Jammu & Kashmir Government (Urdu). Jammu:170–171.

Lawrence WR. (1895). *The Valley of Kashmir*. London: Henry Frowde.

Longstaff TG. (1919) Obituary: Arthur Neve. *Geographical Journal*;54:396–398.

Neve A, Neve EF. (1891) The mortality of ten thousand general surgical operations in the Kashmir Mission Hospital. *Lancet*;II:171.

Neve A. (1892) The cholera in Kashmir. *Medical Missions at Home and Abroad* 1892;iv:164.
Neve A. (1899) *Kashmir, Ladakh and Tibet*. London: Sands & Company: p. ix.
Neve A. (1900). *Picturesque Kashmir*. London: Sands & Company, p. 151.
Neve EF. (1923) Kangri burn cancer. *British Medical Journal*;II:1255–1256.
Neve EF. (1928) *A Crusader in Kashmir*. London: Seeley Services and Co.
Obituary. (1919) Arthur Neve. *British Medical Journal*;1:584.

5 *Kangri*

Environment, Culture, Economics
and Convenience in the Valley of
Kashmir

Rais Akhtar

History of the People and Their Economy

Kashmir is ancient. The earliest known mention is found in the writings of the Greek geographer Hakateeous (549–486 BC) According to him the city of Kaspatyar is situated where the Indus River becomes navigable, the Kashpayors a tribe living on the banks of the Indus. Herodotus (484–431 BC) called Kashmir by the name of Kaspatyros.

A definite reference to Kashmir is found in Ptolemy's geography, around the 2nd century B.C. According to a recent account, "Ptolemy mentions this territory correctly enough between that of the Daradrai or Dards on the Indus. and Kylindrine or the land of the Kulindas on the Hyphasis (Bias River) and eastwards" (Stein, 1977).

The name Kashmir in Sanskrit can be rendered as Ka, water, and Sam-ir, the land from which water has been drained off. This is one theory related to the origin of the valley. According to another theory, Kashmir's real name was Kasheer, from a people named Kasm, who founded what are today the cities of Kashan and Kashga.

A historian one thousand years ago spoke of the inhabitants of Kashmir: "they are particularly anxious about the natural strength of their country and therefore, take always much care to keep a stronghold upon the entrances and roads leading into it" (Rabbani, 1981).

The Kashmiri can turn their hands to anything. They are excellent cultivators, fine gardeners and have a considerable knowledge of horticulture. They can weave excellent woolen cloth and make first-rate baskets. They can build their own houses and can make their own ropes. There is scarcely a thing which they cannot do. As fine craft workers, they may have a few equals in the world but probably none superior. The boatman of Kashmir is as clever as the gondolier of Venice and would emerge safely from the riskiest situations.

The people are persistent, humble as well as intellectual. Still they cling to their land and their traditions, and they probably represent to this day a people historically older than any to be found in Northern India, still associated with the land of their ancestors. Keeping in view the beauty and the productivity of the valley of Kashmir, this is perhaps not so surprising as it would otherwise seem.

According to author Sufi, "Abu Raihan al-Biruni (973–1048 AD) accompanied the expedition of Mehmud against Kashmir probably in 1021 AD, the expedition

DOI: 10.4324/9781003329459-5

being unsuccessful on account of valorous defence by Kashmiris and heavy snowfall (Sufi, 1979)." Sufi further added that the Kashmiri are essentially mystical and imaginative, Huge snowy peaks, flowing silver streams, and sublime solitudes have induced this frame of mind. The imagination of the Kashmiri has given some fine poetry to the world, which, however, has never been fully appreciated for lack of presentation in a suitable form. According to Sufi, the commonest Kashmiri can talk intelligently on most subjects, and has a great aptitude for sarcasm but like other artistic people, is emotional (Sufi, 1974).

About Kangi

Cultural tradition and climatic necessities work together in the continuance of a centuries-old practice of carrying *Kangri*, a portale heat source, under the clothes during the winter months, from about the beginning of November up until late March. The *Kangri* is an earthenware bowl five or six inches in diameter, surmounted by a wicker handle. It is the result of experience spread over several centuries and an improvement over a simple earthenware pot called manna. It contains small quantities of lighted charcoal, sawdust or Chenar leaves.

Toward the end of autumn the valley of Kashmir becomes extremely cold, the trees begin shedding leaves, followed by snow up in the mountains from sometime in mid-December. People wear cloaks (pheron) of fine woolen cloth (pashmina) or the cheaper, rougher pattu or, less commonly, sheepskin (pustin). Rich Kashmiris also have a room or two in their houses with underfloor heating, more familiar to British readers as Turkish bats. The common people use *Kangri*, keeping them close most of the day and night ward off cold. Both rich and poor use *Kangri* outside and inside the house, ad it is offered as the first gesture of welcome to a visitor.

The *Kangri* has become a part of the culture and rich and poor, young and old illiterate and educated all use it. Certainly cold weather is one but not the only reason, as the *Kangri* has not been adopted in far colder areas in the Ladakh region of the Jammu and Kashmir State.

Walter Lawrence whose book The Valley of Kashmir was published in 1895 stated that *Kangri* aids digestion. "A famous native physician was struck with the enormous meals of cold rice and Singhara nut consumed the Kashmiris, but when he saw the *Kangri* he understood that the Kashmiri possessed a remedy against the evil caused by gorging" (Lawrence, 1895).

There are numerous saying and folksongs connected with the *Kangri*:

Kani sana kundalay nivi myani
Kangar
Kya kara chas tsalaan
Kapay viviham kopuy kadhas
Kya kara chas tsalaan
O; which wretch of a woman has stolen my Kangar?
What can I do; I bear the loss; could I catch that wretch, I would tear the hair
   out of her head. What can I do; I bear the loss.

The Kashmiri's love and adoration for *Kangri* may be evident in the following coupler:

> O *Kangri*: O *Kangri*; you are dear to me like a Houri and Fairy; when I take you under my arm, you drive away pain from my heart.

Origin of *Kangri* Use Is Obscure

There is hardly any research carried out on the origin of the *Kangri* tradition. As far back as 1866 a paper appeared in the Indian Medical Gazette written by W. J. Emslie, then Medical Superintendent in Srinagar, about *Kangri* cancer. Emslie did say that the custom of the use of portable braziers for the purpose of warming was not altogether unknown in England, for in "the straw plait districts the children employed in that work are said to carry earthenware or tin pots with them to warm themselves with in winter while engaged at their work". Emslie further stated "the writer saw with his own eyes, during a tour in the north of Italy, the inhabitants of Florence making use of a vessel not very different from the Kashmirian *Kangri*, and for exactly the same purpose" (Elmslie, 1866).

It is quite possible the Kashmiris have learned about *Kangri* from Italian visitors who came to India as craftsmen and some as guests of Mughal Emperors mainly during the 14th and 16th centuries A.D. It is interesting to note that a book written in AD 1148–49 has a reference to *Kangri*. Nineteenth century historian Lawrence further mentions that

> some patriots go so far as to assert that the introduction of the *Kangri*, and its necessary auxiliary the gown, was an act of statecraft on the part of emperor Akbar who wish to tame the brave Kashmiri of the period but others say that the great King Zain-ul-Abidin (AD 1420–70) in his effort to reduce the proud spirit of the Hindus, insisted on the use of *Kangri*....
>
> (Lawrence, 1895)

Kangri Burn Cancer

The very first scientific account of *Kangri* burn cancer in the valley of Kashmir was provided by Emslie; writing in 1866 he observed

> when the weather is extremely cold it is customary for both man and woman while walking abut out of doors, to carry the *Kangri* under their loose woolen gowns, and in close proximity with the bare skin of the abdomen. When indoors, or in a sitting posture, kashmiris place the *Kangri* between their thighs. The fuel consumed between their thighs. The fuel consumed in the *Kangri* is charcoal, and the heat evolved is often considerable.
>
> (Elmslie, 1866)

According to Emslie the cancers on abdomen and thighs are caused by the injurious effects of the heat of *Kangri*, while the use of portable braziers was not injurious

Table 5.1 Categories of heat of *Kangri*

Kashmiri word	English translation	Temperature
Vushun sur	Warm ash	10C.
Tot sur	Hot ash	20C.
Sot nar	Temperate hot	50C.
Josh nar	Very hot	70C.
Tyongal sur	Red hot	200C.

in a similar manner in the case of English and Italians, due to the arrangement of dress in these areas.

Writing in 1923 for the British Medical Journal, E. F. Neve, Surgeon at the Mission Hospital in Srinagar, highlighted the significance of *Kangri* burn cancer. According to him "since the year 1881 operations for epithelioma performed in the Kashmir Mission Hospital have numbered 2,491, of these approximately 2,000 or 84% were *Kangri* cancer" (Neve, 1923). It is also common to see the skin burn on the abdomen and thighs. In many cases clothes are also burnt. Despite that the cultural tradition is so strong and deep rooted that the practice of *Kangri* survives.

According to findings of a 1975 study, an "overwhelming majority of cases of deformities resulting from burns reported in Kashmir, are attributed to the *Kangri*". Another investigation revealed that

> a group of doctors, who made statistical analysis of 1,500 cases at the Shri Maharaja Hari Singh Hospital, Srinagar, stated in the report that *Kangri* was the cause of post-burn contracture deformities in 95% of the cases. The doctors further reported that all the patients came from law and middle income groups.
>
> (Sathu, 1975)

A. N. Raina in his book Geography of Jammu & Kashmir throws light on the heat of *Kangri*. He classifies it in five categories (Raina, 1971) (Table 5.1):

A Changing Way of Life

It is still true that *Kangri* is widely used by both poor and rich not only in Kashmir but in different parts of the country, including comparatively warmer state of Kerala since the *Kangri* tradition is part of the culture: nevertheless the lifestyle of Kashmiris has changed considerably. Now most of them wear two or three woolen garments inside the main *pheron*. With the spread of education and involvement in different types of jobs, contact hours with the *Kangri* have been reduced. The recent initiative by the Central and State Governments to provide more opportunities for work during winter months will improve people's prosperity and will minimize the use of *Kangri*. This will go a long way towards eradicating *Kangri* burn cancer in the Valley of Kashmir. The average five or six cases annually (personal discussion with Prof. Sanyal and Dr. Barua) of *Kangri* burn cancer in recent years represents

Figure 5.1 Kangri

substantial progress compared with the figures quoted by E. F. Neve. According to a recent research by Imtiaz Wani, "Kangri cancer still occurs in patients who live in remote areas of Kashmir where there is a chilling cold in winter months and have no modern and alternative cheap means of warming other than the Kangri" (Wani, 2009). Based on data between 2008 and 2018, analysis of data of skin burn cancer suggest that Kangri was the most common etiological agent of epileptic burns in 99 (63%) patients. Eighty-four (53%) had sustained full-thickness burns and total body surface area involved was up to 5% only (Baba et al., 2019). Recent report suggests that there is a concern over the rising number of Kangri cancer cases in the region, Some 400 patients with skin cancer including the Kangri burn cancer were registered at the Sher-I-Kashmir Institute of Medical Sciences, Srinagar, in the last five years (G.K., 2022) (Figure 5.1).

References

Baba, P.U.F, Sharma, S.K., and Wani, A.H. (2019) Epileptic burn injuries in Kashmir valley: Is "Kangri" a boon or bane? *Indian Journal of Burns,* Vol.27, Issue 1, pp.95–101.

Elmslie, W.J. (1866) Etiology of epithelioma among Kashmiris, *Indian Journal of Medical Gazette*, November 1, Vol.1, pp.324–326.

G.K. (2022) Kangri cancer cases on the rise in Kashmir, *Greater Kashmir*, April 29.

Lawrence, W.R. (1895) The Valley of Kashmir, London, H. Frowde, p.251 (Asian Educational Services, 2005).

Neve, E.F. (1923) Kangri- burn cancer, *British Medical Journal,* Vol.11, pp.1255–1256.

Rabbani, G.M. (1981) *Ancient Kashmir: A Historical Perspective*, Gulshan Publishan, Srinagar, p.2.

Raina, A.N. (1971) Geography *of Jammu & Kashmir*, National Book Trust, New Delhi, p.100.

Sathu, J.N. (1975) Kangri, *The Consumer Gazette* (Independence Day Kashmir Special), August 10, p.28.

Stein, M.A. (1977) The Ancient Geography of Kashmir, Journal of Asiatic Society of Begal, Vol. LXVIII, Part I Extra Number 2, 1899. Published in Book form by Indological Book Corporation, Patna, p.9.

Sufi, G.M.D. (1974) *A History of Kashmir*, Vol.1, Light & Life Publishers, New Delhi, p.21.

Sufi, G.M.D. (1979) *Islamic Culture in Kashmir*, Light & Life Publishers, New Delhi, p.10.

Wani, I. (2009) Kangri cancer. *Surgery*, Vol.147, p.586–588. 10.1016/j.surg.2009.10.025.

6 Environment and Health in Kashmir during 19th Century

Rais Akhtar

Introductory

Historical Medical Geography is a fascinating area of research which highlights the pattern of environmental conditions, people's adjustment with the environment, and the incidence of diseases as a result of man's interaction with the environment.

Kashmir had undergone various phases of historical/political development since the beginning of 19th century. As a result of the socio-cultural, political and economic conditions, varied in different period of time leading to wide variation in diseases and health care framework in Kashmir.

Geographically the valley of Kashmir with clayey loamy soils, is nestled in the young folded mountain ranges of the Himalayas. Its water features are the principal components in its scenic beauty. The climate is temperate, being neither too hot nor too cold but in winter the temperature descends several degrees below freezing point.

Socio-Economic Conditions

However, poor economic conditions and unsanitary and unhygienic living conditions present conducive environment for the occurrence of diseases. A report dated 10 December 1861 from the Deputy Commissioner on special duty at Kashmir to Secretary Government of Punjab is a testimony of the poor economic conditions of the people in Kashmir. The report states, "The People in Kashmir are wretchedly poor, and in any other country their state would be almost one of starvation and famines" (Report, 1861). Under such socio-economic circumstances, people were not able to maintain a hygienic environment. Dr. William Jackson Elmslie who arrived in Srinagar as a missionary doctor in 1865, described in his letters the unhygienic practices of the people which result in the widespread incidence of skin diseases. In one of the letters he writes,

> But there are other things in Kashmir which most terribly detract from its pleasure as a place of residence. The dirt is beyond description. Who can tell what Kashmir smells are? Not the odour of roses, such as one has expected to fill the air; but oh! such, that the dirtiest of London Courts is sweeter than the cleanest of Kashmir villages. The clothes, too, of the people are filthy;

DOI: 10.4324/9781003329459-6

not that the filth shows much, for all their garments are of grey wool, which is a most perfect concealer of dirt; but not a few of their diseases are the result of their uncleanliness, and how often I have almost shrunk away from them, as, in my dispensary, while I have been examining a patient. I have seen the lice crawling on his clothes and his fleas skipping over to me. Of course, if you can avoid all intercourse with the natives, than dirt is not such a continual source of annoyance, but to us it was a daily trouble.

(Elmslie, 1875)

Walter Lawrence, who was then Settlement Commissioner in Kashmir also noted in 1895, "The clothes of the villagers are simple and extremely mean in appearance" (Lawrence, 1895).

As a result of socio-cultural practices including poor economic condition and insanitary environment, a number of infectious diseases and Kangriburn cancer were widely prevalent in Kashmir. It would be interesting to note that Kangri-burn cancer is region-specific and was being reported only in Kashmir.

Kangri-Burn Cancer

Cultural tradition and climatic necessities work together in the continuance of a centuries-old practice of carrying KANGRI, a portable heat source, under the clothes during winter months, from about beginning of November up until late March. The KANGRI is an earthenware bowl five or six inches in diameter, surrounded by willow basket work and surmounted by a wicker handle. It contains small quantities of lighted charcoal, sawdust or chenar leaves (Akhtar, 1992). The very first scientific account of KANGRI burn cancer in the valley of Kashmir was provided by the first medical missionary doctor, Dr. W. J. Elmslie who came to Srinagar in 1865. He observed, "when the weather is extremely cold it is customary for both man and woman while walking about out of doors, to carry the KANGRI under their loose woollen gowns, and in close proximity with the bare of skin of the abdomen. When indoors, or in a sitting posture. Kashmiris place the KANGRI between their thighs. The fuel consumed in the KANGRI is charcoal and the heat evolved is often considerable" (Elmslie, 1866). According to Elmslie the cancers of the abdomen and thighs are caused by the injurious effects of the heat of the KANGRI (Akhtar, 1992).

Writing in 1923 for the British Medical Journal, E.F. Neve, Surgeon at the Mission Hospital in Srinagar highlighted the significance of Kangri-burn cancer. According to Neve, "Since the year 1,881 operations for epithelioma performed in the Kashmir Mission Hospital have numbered 2,491, of these 2,000 or 84% were Kangri cancer" (Neve, 1923).

Epidemics in Kashmir

Besides floods, famines, fires and earthquakes, Kashmir was affected by a number of epidemics, such a cholera, plague and smallpox during the 19th century. But according to Lawrence the great scourage of Kashmir was cholera. It erupted in different epidemic waves in the 19th century, which caused sudden and widespread mortality in Kashmir.

Sudden outbreaks of cholera in Kashmir during the 19th century have been described by many writers and physicians. The first recorded cholera epidemic broke out in Kashmir in December 1857. The epidemic lasted upto the middle of August 1858. The outbreak of cholera was reported again in 1867. Several thousand deaths were reported. In 1872 cholera again broke out in Kashmir in June. When famine is combined with cholera as was the case during 1877–79, it caused havoc among the people of Kashmir.

One important aspect of geomorphic characteristics of the valley is the pattern of soil distribution. Writing in early 1890s, Walter Lawrence noted, "it is a curious fact, which I noticed in tours, that villages on the Karewa plateaus seemed free from cholera, and that the disease was most rampant in the alluvial parts of the valley" (Lawrence, 1895). An examination of geoecology of cholera reveals among other factors, that apart from man (as both pathogens and geogens), there are environmental factors which facilitate the survival and spread of cholera. Such geographical factors include-annual temperatures of around 37 °C and prolonged dry spells. Moisture is another important factor which favours the outbreaks of cholera. Cholera vibrios thrive best in warm alkaline media (with pH value between 9 and 9.6), while acidic media inhibit their growth and survival. The Karewa soils are devoid of vegetal cover and are deficient in organic matter. The moisture-retaining capacity of the soil is poor as the upper layer has a high sand content. Therefore, the characteristics of soils may explain the statement of Walter Lawrence regarding the absence of cholera in the villages on the Karewa soils of the valley of Kashmir (Akhtar, 1992).

Fortunately, however, Kashmir remained free from cholera from 1880 to 88. In 1888, there was acute shortage of pure drinking water in Srinagar City. From the point of view of sanitation, the city was like a powder magazine waiting for a spark, and this was supplied in the following spring. The cholera broke out in April in a virulent form. About 10,000 human lives were lost within two months. According to Sufi, a historian, the death toll was 18,000. Cholera again visited the valley in 1892. The epidemic this time was a disastrous one. It ranged for about four months from May to August and rendered the people panic-stricken. The mortality rate was so high that exact registration was quite difficult, particularly in rural areas. In Srinagar the number of deaths was so large that during day time the corpses were burnt by the Hindus and buried by the Muslims according to their religious practices respectively, but in the night they just threw dead bodies into the river (Dev, 1983).

Walter Lawrence quoted Surgeon Colonel Harvey, who visited Srinagar during the epidemic of 1892, writes, "It is not too much to say that the inhabitants eat filth, drink filth, breathe filth, sleep on it, and are steeped in it and surrounded by it on every side" (Lawrence, 1895).

After the cholera epidemic of 1892, Dr. Mitra, the then Chief Medical Officer of Kashmir, took note of the observation of Col. Harvey and started vigorous measures for better sanitation of Srinagar city and other towns. Roads and drains have been made, a supply of pure water has been started. These measures of public sanitation have greatly influenced the habits of the people, and thus the cause of both public and private hygiene had improved with rapid strides.

This was followed by vaccination which became very popular in Srinagar. Dr. Mitra who was incharge of the Vaccination Department says, "opposition, which is sometimes very strong, is met with mild persuation and patient submission, and in this way alone, I have succeeded in overcoming what appeared to me once insuperable difficulty" (Annual, 1895–96). According to E.F. Neve, "until the introduction of general vaccination, practically the whole population of Kashmir contracted smallpox in childhood" (Annual, 1895–96).

The efforts made by Dr. Mitra regarding sanitary measures as well as vaccination, resulted in total eradication of cholera and considerable decline in smallpox incidence in Kashmir during 1895–96. Following Table 6.1 shows that during the 1895–96 ulcers, syphilis, malarial fever, eye diseases and skin diseases dominated the health problems in Kashmir. It should be noted here that malarial fever was confined mainly in Karnah tehsil of Muzafferabad district of Kashmir. Diseases of the digestive system, worms, and respiratory diseases were also widely prevalent. Being a mountainous region with deficient iodine in soil and water some 581 cases of goitre were reported during 1895–96 (Figure 6.1).

During the same period i.e. 1895–96 a large number of surgical operations were performed, such as removal of tumour and operations for cataract and deformities.

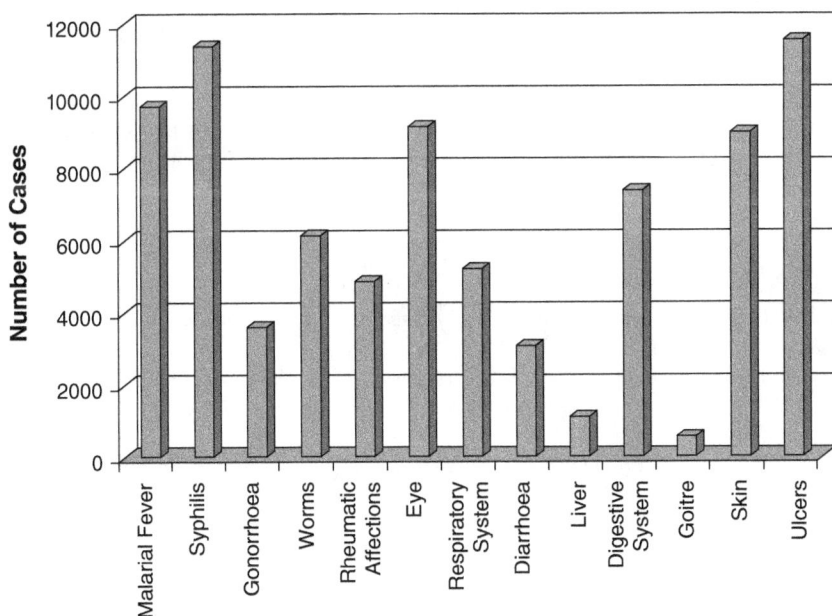

Figure 6.1 Principal diseases in Kashmir, 1895–96 (By the Author).

Table 6.1 Principal diseases in Kashmir during 1895–96

Diseases	Number of cases
Malarial fever	9.713
Syphilis	11.394
Gonorrhoea	3.608
Worms	6.128
Rheumatic affections	4.892
Eye diseases	9.165
Diseases of respiratory system	5.250
Diarrhoea	3.059
Diseases of liver	1.143
Diseases of digestive system	7.406
Goitre	581
Skin	9.010
Ulcers	11–551

Source: Annual Administration Report of J&K, 1895–96, Jammu, p. 92.

Thirty-one surgical operations were performed for the removal of epithelioma and Kangri ulcers during 1895–96.

Conclusion

The study presents an account of environmental conditions in relation to the prevailing diseases in Kashmir during 19th century. The paper discusses the man–environment interaction in the framework of changing socio-cultural, economic and political environment in the context of changing disease scenario in Kashmir. Future research into historical medical geography of environment and diseases in a region should focus both on environmental conditions in particular regions and on the relationships between the interaction of the social, cultural, economic, political and physical environment and diseases of the region.

References

Akhtar,R. (1992) KANGRI: *Traditional personalized central heating in the valley of Kashmir,* in "Focus" vol. 42. *2,* p. 27.

Annual Administration Report of-J&K Government, 1895-96, Jammu, pp. 95–96.

Dev, J.S. (1983) *Natural Calamities of Jammu and Kashmir (*1993), New Delhi, Ariana Publishing House, p. 122.

Elmslie,W.J. (1866) *Etiology of epithelioma among the Kashmiris,* in "Indian Medical Gazzette" vol. 1, pp. 324–326.

Elmslie,W.J. (1875) *Seedtime in Kashmir: A Memoir,* London, James Nisbet & Co., p. 79.

Lawrence,W.R. (1895) *The Valley of Kashmir,* London, Henry Frowde.

Neve,E.F. (1923), *Kangri-burn cancer,* in "British Medical Journal" vol. 11, pp. 1255–1256.

Report from Dy. Commissioner on Special Duty at Cashmere to Government of Punjab, dated 10th December 1861. Akhtar.

7 Changing Diseases Ecology of Leh District

Contemporary Scenario and Historical Perspective

Rais Akhtar

Geographical Location

The town of Leh is situated on an alluvial fan. To the east of Leh on either bank of Indus is an alluvial plain stretching over a distance of 30 Kms. This is the most fertile levelled land inhabiting several rural settlement of the Ladakh region. Although its physical isolation has prevented rapid change, the expansion of tourism due to disturbed conditions in Kashmir since 1990 has provided a lot of inputs for changing lifestyle of the people on pattern of western style development. Despite this contemporary influence, the culture of Ladakh which is historically closely linked to that of Tibet is generally well preserved (Norboo et al., 1991).

Geo-ecology of Health

There is a strong linkage between geo-environmental conditions and an epidemiological scenario in Ladakh. The Indus valley in Ladakh is subject to periodic loss derived dust storms. For some years local doctors have been concerned at the frequency of chest disease in some village of the Indus valley in Ladakh. Table 7.1 shows the number of cases of various diseases reported in Leh District (Figure 7.1).

Respiratory tract infection (RTI) dominant health problem in the Leh region. The highest cases of respiratory illness were found in the village of Chuchat Shamma which lies at an altitude of 3,000 meters (11,200 feet) beside the river Indus, and almost 15 Kms from the town of leh. This region lies in the rain shadow of the main Himalayan range known as Ladakh ranges. Indus, though flowing through gorge has numerous alluvial fans and river terraces. RTI is followed by dental diseases, diarrhoael diseases, hyperacidity cases, ophthalmic diseases and skin diseases. The geo-ecology of the health hazard in Leh must be understood in this regard. Major health hazard to health in Leh is the transport and transformation from primary dust source (loess) to airborne dust to which local communities are exposed. The population is exposed to silica-rich particulates with a substantial respirable size fraction. Studies carried out in the region note signs of pneumoconiosis in X-rays of some villagers in a region with no mines or dusty industries. Two important surveys of silicosis, carried out in the surrounding areas of Leh, found that, in one relatively highly exposed village, the prevalence reached 45% in a random sample

DOI: 10.4324/9781003329459-7

Table 7.1 Incidence of diseases in Leh District

1 Respiratory Tract Infection (RTI)	11,957
2 Dental cases	9,123
3 Acute Diarrhea cases	6,828
4 Hyperacidity cases	3,397
5 Aphthalmic cases (other than cataract)	2,670
6 Skin problem	1,861
7 ENT cases	1,369
8 Hypertension	1,042
9 Mumps	0,333
10 T.B cases	0,073
11 Chiken pox	0,076
12 Leprosy	0,003
13 STD	0001

Source: Data pertains to the period 2000–01 (upto ending July) Chief Medical Officer, Leh).

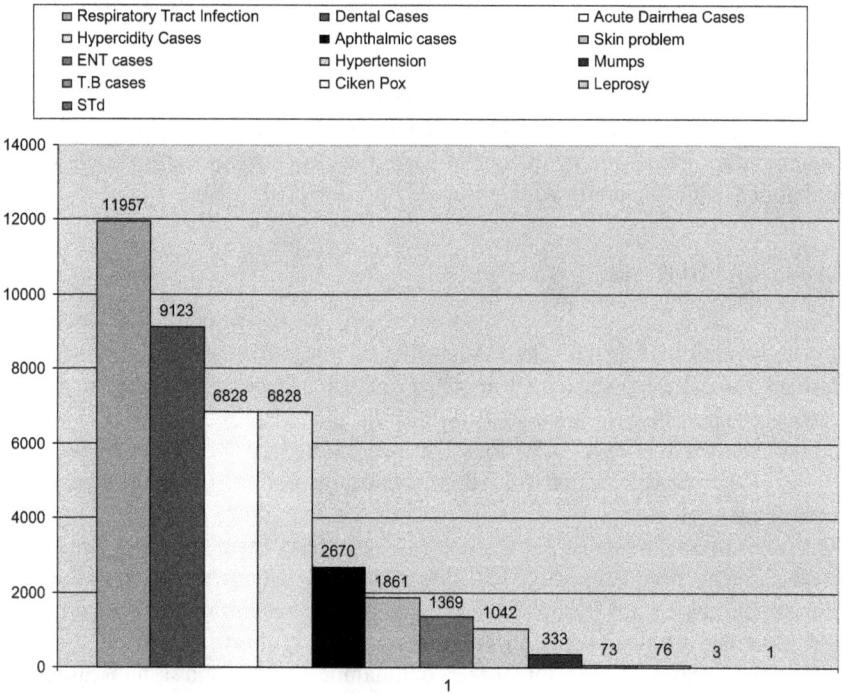

Figure 7.1 Incidence of diseases in Leh district.
Source: CMO, Leh, 2000–01.

of 150 adults aged 30 and over, including cases of progressive massive fibrosis in the villages of Chuchat Shamma and Stok. The villages are exposed to frequent dust storms. Chest radiographs of villagers aged 50–62, showed varying grades of silicosis, compared with 3 of 13 men and 7 of 11 women in Stok, which lies 300 metres higher and is exposed to fewer dust storms. The study concludes that the difference in prevalence of silicosis between the two villages was significant, as was the differences between men and women. The findings of the study support the evidence that silicosis may develop when people are exposed to desert dust. The study suggests that silicosis is widespread among older people exposed to environmental dust and that it may result in advanced fibrotic lung disease associated with disability. Women, who are more heavily exposed to dust in the course of their work, appear to be more commonly affected then men (Norboo et al., 1991). The investigations reveals that the reduction of exposure in this community would be difficult to achieve. This is because of the fact that geo-ecology of the region-types of rocks/soils which are rich in silica and the frequency of dust storm will continue to affect the population in the Ladakh region. It is interesting to note that a different kind of dust storm causing havoc in northern China affecting the food production and according to the study may lead to starvation and malnutrition. The

> Gigantic dust clouds swirling over China are threatening the world's most populous country with the first- ever ecological meltdown, experts warns. The clouds – which stretch for thousands of miles over Asia and have even reached across the pacific to North America- are rising from a rapidly growing dust bowl in northern China that far out-strips the notorious one in the United States in the 1930s. The clouds sweep up millions of tons of precious topsoil from Chinese fields and pastures.
>
> (Lean, 2003)

Thus the problem related to environmental dust is a continuous problem in this region of Asia. It has therefore been suggested that the reduction of dust exposure in this community would be difficult to achieve. This emphasis the need to prevent respiratory diseases caused by other factors, such as cigarette smoking and smoke pollution from fires in the home.

Climate Change Phenomenon

Although one of the most experienced physicians of Leh Dr.Norboo[1] suggests that the occurrence of dust storm, in this region of Ladakh is a regular phenomenon, and not an unusual thing for this region; however, the author of this paper could not find in the literature the geography and history on Ladakh, any example of the occurrence of dust storm in Ladakh. Most important example, that too by an American geographer-Ellsworth Huntington, who visited Kashmir and Ladakh in 1905, tells us about ecology of Ladakh.

"Arid, inhospitable, and rugged as Ladakh may be, its clear air, bracing climate, and splendid scenary make the traveler long to return to it. The stony villages and ugly people have a pecular charm" (Huntington, p. 58). While traveling at Puski and Zanguya in Khirgiztan, Huntington does mention dust storms;

> There was nothing to look at except pebbles, wonderfully smoothed and faceted by wind-blown sand, or dense colums of whirling dust, thirty or forty feet in diameter at the base and rising to a height of hundreds or thousands

of feet, where they spread out after the manner of thunder clouds. Twice I counted between twenty and thirty dust-whirls visible at one time, and there were always at least eight or ten. It was evident that, even if there were no wind, the air in summer would be full of dust continually.

(Huntington, pp. 148–149)

The above discussion suggests that dust storms in the Ladakh region are a recent phenomenon. There is good scope to investigate the climate change phenomenon on impact the occurrence of dust storms, and as a result of this respiratory diseases are wide spread in this region.

Socio-Economic Impact

Based on the survey carried out by the author in the villages around Leh, it was noted that environmental dust has not only caused health problems, but also influenced the socio-economic framework. As a result of silicosis women hardly reach the age of 50. Besides boys from outside from this region. Chuchat, Stakna, Thikse and Shey would not prefer to marry girls from these villages because of high prevalence of silicosis. Besides,

women, who are more heavily exposed to dust in course of their work, appear to be more commonly affected than men. (Norboo et al., 1991). The region is endemic for silicosis. The yield of various crops such as wheat, barley, peas and vegetables has declined. Although it has not been studied so far but a trace element pathway analysis might throw light if there is any link between the silica content in the soils-food grown on such soils and the resultant impact on health since, "environment deficiencies or overloads of trace elements are mediated through the consumption of drinking water and locally grown food-stuffs, reflecting customary dietary patterns.

(Hunter and Akhtar, 1991)

In view of the seriousness of the environmental hazards in the region, some researchers have suggested to shift those villages from this endemic region. This region is prove to the high wind velocity.

The high frequency of wind blow causes soil erosion and environment dust pollution. The problem of environmental dust has also been aggravated after the construction of the road and by the passing vehicles of the road and the passing vehicles. This is an area dominated by Muslim population, and they are also heavy smoker. It is just not understandable as to why in such an silicosis prove area two stone crushers and one cement factory were established. Several demonstrations were also held against the construction of the road. Efforts are being made to popularize the smokeless stove, and the use of mask against dust among the population in the region.

Changing Disease Ecology: Historical Perspective

In my quest to understand the disease ecology of Ladakh during the later part of 19th century, I found out in the Welcome Library, London, one short paper entitled, "Notes on Ladakh in 1867, written by Assistant-Surgeon Henry Cayley, in 1868 Indian Medical Gazette". This paper, apart from discussing the pattern of diseases in Ladakh (Leh and surrounding villages), also highlights the politics of health care availability in Ladakh. There used to be an active opposition of the introduction of western medicine by a Hakim (Unani practitioner) from Kashmir, the Kashmiri officials of Maharaja in Ladakh would secretly place hurdles before patients leading to decline in patient attendance at the allopathic dispensary Cayley writes,

> ..... owing to the obstruction secretly thrown in the way by the Cashmere officials, the attendance almost ceased; but after a short time I managed to put a stop to all active opposition, and the attendance of sick of all classes, both from Leh and its neighbourhood, and from distant places, at once revived at the same time an opposition dispensary was opened under the charge of a Hakim from Kashmir, and for a time the patients on their way to me were forcibly stopped and taken there for treatment, but as soon as this system was abandoned, the attendance at the Maharajah's Dispensary entirely ceased; for people of Ladakh do not believe that any good thing can come out of Cashmere.
>
> (Cayley, 1868)

It is interesting to note that medicine was free from the western dispensary, however the Hakim and other indigenous "medicine Men" (most probably Amchis)[2] used to charge fee from the poor patients. There was a local saying about a Doctor:

When the care complete, he seeks his fee;
The Devil seems less terrible than he

Amchi is presently the most popular traditional (indigenous) system of Medicine in Ladakh.

The Table 7.2 shows the diseases reported during July and August 1867 at the western Dispensary.

It is evident from the Table 7.2 that the disease pattern in Ladakh in 1867 was quite different from the contemporary disease scenario. Fever, dyspepsia, ophthalmic, rheumatism, caries of teeth, syphilis were dominant diseases. Anaemia, scabies, hepatitis, frostbite, sinus and eczema were insignificant. Six cases of carcinoma were also reported. According to Cayley fever appears to form a larger proportion of the sickness due to indigestion, cold, exposure to the skin whilst at work, standing in cold water, etc. Cayley further adds that except in pilgrims and merchants, and others coming from the plains, no fever case similar to malarious fever was reported. Describing the favourable ecological condition for malaria, Cayley says- "… I hardly believe the disease to exist in spite of the whole of the land in the villages being almost constantly under water and exposed to a powerful sun----" (Cayley, 1868) (Figure 7.2)

Table 7.2 Incidence of diseases in Leh, 1867

Diseases	Number of cases	Diseases	Number of cases
Fever	66	Constipation	2
Dyspepsia	61	Lepra	2
Ophthalmia	45	Anaemia	1
Rheumatism	38	Scabies	1
Caries of teeth	32	Hepatitis	1
Syphilis	25	Frostbite	1
Neuralgia	17	Sinus	1
Ulcer	16	Eczema	1
Bronchitis	16	Scrofula	1
Colie	10	Orchitis	3
Gonorrhea	8	Dysentery	3
Cataract	7	Laryngitis	3
Scorbutus	7	Diarrhoea	2
Carcinoma	6	Cephaloea	3
Entropion	4	Fattytumours	2
Paralysis	4	Tonsillitis	2

Source: Cayley, H, 1868.

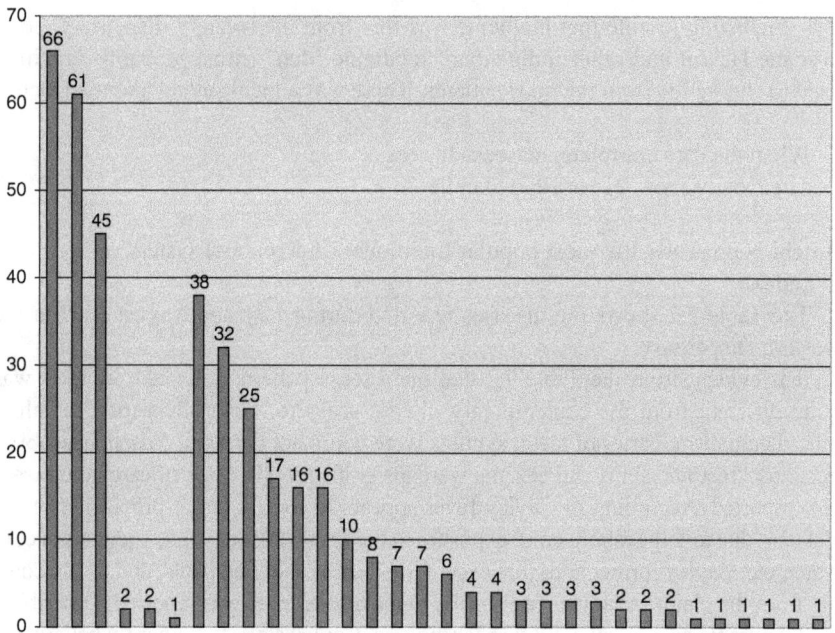

Figure 7.2 Incidence of diseases in Leh, 1867.
Source: Cayley, H. 1868 Index of Diseases.

Arthur Neve's Description of Socio-economic Ecology of Ladakh

In his letter to the Christian Missionary Society, Arther Neve described the sufferings of Ladakhi people due to cataract, difficult physical accessibility and the difficult conditions under which Neve carried out operations while holidaying in the Ladakh region in 1898.

Neve Writes

in some of the villages I was able to stop and treat the sick, extracting a few cataracts, as these people would never have a chance of going to hospital, though I tried to persuade some to go to Leh, where there is a surgeon of the Moravian Mission. But with a pass 17,000 feet high to cross, these poor blind people could not go. I met indeed one man, a Buddhist priest, who had gone in 1896 and received sight at the hand of one of us, but as a priest he had special facilities for traveling who would help blind women over the pass, and procure yaks for them to ride on.

Once when it was dusk, in a village in Ladakh, beyond the swift bridgeless Shyok, three poor creatures arrived in my camp. At first I refused to operate and told them to go to Leh. We were to start long before day break and cross the river by ferry, then do a long march up the mountain, but their importunity prevailed, and I said if they would be at the river I would see what could be done. At down it was both windy and rainy. At the ferry these poor women had slept with no food but a little raw dough. We got into the boat and were swiftly swept down among the leaping wave, and landed a quarter of a mile down the other side. The ferry boat returned for the rest of our party. I had my box of instruments, but how I sterilize them and how should I light a fire?

I told the Ladakhis and they tried to strike sparks with flint and steel but the tinder seemed moist. One of them then produced a little gunpowder and placed it on a stone, tore off a rag from his shirt and fraying it out laid it by the powder, then with flint and steel ignited it. Then a cooking pot was produced and soon water was boiling. What an anachronism between the aseptic surgery aimed at and the primeval method of fire production! While the instruments were being boiled I cleaned the eyes and instilled cocaine, then kneeling in the sand removed the three cataracts, completing the operations, just before a gust of wind came. laden with dust and grit, which would have put a stop to my work. The gratitude of the people knew no bounds.

(Neve, 1898, p. 14)

Diarrhoea and Dysentery are almost unknown Couple of cases of Diarrhoea and dysentery were reported of people coming from Kashmir and Kullu now in Himachal Pradesh. Thus it seems that the production of contamination of water was almost negligible. Similarly cholera according to Cayley,

has not reached Ladakh, though this year it has been raging in cashmere, and come very close to the frontier, but it never surmounted the pass between the two countries; and as there was constant intercommunication, I can only suppose that the poison of the disease cannot produce its effects at an altitude of 10,000 feet above the sea.

(Cayley, 1868)

The statement provides a scenario of the process of cholera diffusion and difficult physical barrier of Zojila pass between Kashmir and Ladakh which hindered the diffusion of cholera wave from Kashmir. It is significant to note that Cayley referred cholera as poison of the disease. It would not be out of context to mention that in 1849 John Snow published a small pamphlet "on the mode of communication of cholera" where he proposed that the "Cholera Poison" reproduced in the human body and was spread through the contamination of food and water. It seems Cayley was very well aware about John Snow's work.

Conclusion

The study of changing scenario disease ecology in Ladakh throws light on the pattern of Socio-economic and cultural practices in relation to the occurrence of disease in Ladakh, that have ravaged the population during the middle of 19th century, and those which occur at present. The paper emphasizes the role of climate change in causing dust storms leading to the higher incidence of chest diseases as a result of RTI, although a more intensive study should be carried out on this aspect. Since the objective of the present study is to present a comparative disease scenario between 1867 and 2001, and to highlight geographical explanation on the pattern of diseases.

Note

1 Personal Communication with Dr. T. Narboo, Leh. (June, 2003).

References

Akhtar, R. (1991) *Environment and Health: Themes in Medical Geography*, Ashish Publications, New Delhi.
Cayley, H. (1868) Notes on Ladakh in 1867, *Indian Medical Gazette*, January 1, p.3.
Hunter, J.M. and Akhtar, R. (1991) The challenge of medical geography, In R. Akhtar (ed.) *Environment and Health*: Themes in Medical Geography, APH, New Delhi, p.30.
Lean, G. (2003) Huge dust cloud threatens Asia, *Independent Sunday*, 26th January, London.
Neve, A. (1898) Christian medical missionary, *Annual Letters*, University of Birmingham Archives, p.14.
Norboo, T. et al. (1991) Domestic pollution and respiratory illness in a Himalayan village, *International Journal of Epidemiology*, vol. 20, no. 3, p.749.

8 Traditional Medical Therapy in Rural Ladakh – A Regional Analysis

G. M. Rather

Introduction

Indigenous medical beliefs and practices have always been an integral part of many cultures (Bhardwaj, 1985) as cultural aspects play an important role in health care as variables like ethnicity, language, religion, political, economic aspects have witnessed different health care systems (Gesler, 1991) and even cultural beliefs and habits effects utilization of health care facilities (Chandri, 2002). Every culture irrespective of its simplicity or complexity has its own beliefs and practices concerning disease and evolves its own system of medicine and this medical system is known as traditional medical system (Fonaroff and Fonaroff, 1985).

Ladakh, the northern most part of India with an area of 96,701 sq.km in the Trans-Himalayan region of India lies between 32° 15 / to 35° 55 / north latitude and 75° 15 / to 80° 15 / east longitude Located at an average altitude of 3,500 m, the region experiences arid and cold climate. Mean maximum temperature is 24°C, and mean minimum temperature is -20°C. Winters are of long duration (October–April) (Husain, 1998) The settlements are scattered in remote inaccessible areas where the allopathic medical system facilities are meager.

The region is rich in ethnic folklore, culture, and heritage and is characterized by its own geographical peculiarities (Sagwal, 1991). The region has a rich biodiversity of medicinal plants and most of the people especially in the rural areas still rely on medicinal plants for their treatment during ill health. The traditional medicinal system prevalent in the area is commonly known as Amchi medical system – an off shoot of Tibetan medical system (Choedak, 2000).

Traditional medical system , one of the approaches (Paul, 1985) and major subfield of Medical Geography (Hunter and Akhtar, 1991), has received attention only after 1970 (Good, 1977) but a very good contribution has been made in this field by medical geographers and some notable contributions are; Traditional healers in rural Hausland by (Stock, 1981); Sidah medical system in Madras by (Ramesh & Hyma, 1985); Indigenous practitioners of Sidah medicine in south India by (Hyma, Ramesh and Srivastava, 1986) health related beliefs and attitudes of Hausa people of Nigeria by (Stock, 1980). Some other remarkable contributions are that of Phillips (1984); Ramesh and Hyma (1981); Warren (1982); Akhtar and Rather (1999).

DOI: 10.4324/9781003329459-8

The present research in the same direction was carried out to investigate the role of traditional knowledge in the medical therapy of Ladakh.

Data Base and Methodology

Data Base

The present study is mainly based on primary data and partly on secondary data. Secondary data was collected from census handbooks, books and Journals, while as primary data was collected through field survey.

Methodology

Selection of Sample

The study area was delineated into six physiographic regions (six Physiographic Divisions are the valleys of which the Ladakh comprises of – Nobra, Leh, Chang-thong, Zaskar, Suru and Drass. Stratified Random Sampling technique was used for the selection of *18* sample villages (15% of total revenue villages, three from each region but all falling in rural areas. For household survey, sample size of 200 households comprising 20% households of each sample village were selected by employing stratified random sampling technique.

Sample Survey

An intensive and effective household survey of 200 households in rural areas of Ladakh was conducted with the help of a structured questionnaire containing questions regarding socio-economic variables of households, utilization and preference of traditional medical system and the source of knowledge behind the utilization of this medical system, perception of disease and management of different types of disease in different regions of Ladakh.

28 Am chi practitioners comprising 10% Sample size were selected by using stratified random sampling technique. Interviews were carried out with Am chi practitioners regarding practice of Am chi medical system and their socio-economic aspects. Source of traditional medical system, herbs and their distribution in the area, part of the herb used in different treatments and their satisfaction with the job were the main aspects of discussion with Amchi. Altitude of different locations of herbs was noted from the topo-sheets of the study area.

Data Analysis

The data collected through household survey and interviews with Am chi was tabulated, analyzed and interpreted. Statistical relationship measures like correlation, regression and coefficient of determination were used to find out the association between socio-economic variables and traditional medical system.

Table 8.1 Utilization and preference of medical systems and source of knowledge in Ladakh

Physiographic divisions	No. of households surveyed.	Medical system utilized (no. of households with %age to total)			Preference of Amchi medical system	Sources of Knowledge for tilization of Amchi medicine. Traditional literature	
		Allopathic	Amchi	Both		Amchi medicine.	Traditional literature
Nobra	37	15 (40.52)	17 (45.97)	5 (13.51)	23 (62.16)	8 (47.05)	9 (52.95)
Leh	40	21 (52.50)	15 (37.50)	4 (10.00)	24 (60.00)	8 (53.33)	7 (46.66)
Changthong	38	12 (31.58)	23 (60.53)	3 (7.89)	30 (78.94)	12 (52.17)	11 (47.82)
Total for Leh District	**115**	**48 (41.74)**	**55 (47.83)**	**12 (10.43)**	**87 (75.65)**	**28 (50.91)**	**27 (49.09)**
Zanskar	28	10 (35.72)	14 (50.00)	4 (14.28)	19 (67.87)	11 (78.57)	3 (21.43)
Suru	34	18 (52.94)	13 (38.25)	3 (8.81)	21 (61.76)	7 (53.85)	6 (46.15)
Drass	23	10 (43.48)	10 (43.48)	3 (13.04)	12 (52.17)	6 (60.00)	4 (40.00)
Total for Kargil District	**85**	**38 (40.70)**	**37 (43.54)**	**10 (11.76)**	**52 (61.17)**	**24 (64.86)**	**13 (35.14)**
Total for Ladakh Division	**200**	**86 (43.80)**	**92 (46.00)**	**22 (11.00)**	**129 (64.50)**	**52 (56.52)**	**40 (43.48)**

Source: Based on data obtained from field survey - 2015.

Table 8.2 Relationship between Amchi medical system and Socio-economic variables

Socio-economic attributes	Regression equation	Co-efficient of determination r^2	Co-efficient of correlation (r)
Levels of income	Y=72.32+0.881x	+0.788	+0.888
Level of education	Y=97.88+0.521x	+0.273	+0.523
Age	Y=39+0.756x	+0.927	+0.963

Source: Calculated from data obtained from field survey - 2015

Results

Utilization and Preference of Medical Systems

Analysis of data reveals that out of 200 households surveyed in Ladakh division 46% were found utilizing traditional medical system and 64.50% were found giving preference to this medical system. Utilization of traditional medical system was 47.83% in district Leh and only 43.54% in district Kargil. The preference of Amchi medical systems was found very high in district Leh (75.65) and only 61.17% in district Kargil. There is a very good regional variation in both utilization and preference of traditional medical system. Sources of knowledge for utilization of traditional medical system are both traditional as well as literature. Near about 56.52% were having traditional knowledge of using traditional medical system and 43.48% have learned the utility of using traditional medical system from the literature (Table 8.1) It was found that there is very good relationship between utilization of Amchi medical system and socio-economic variables like age, income and levels of education (Table 8.2).

Socio- Economic Characteristics of Traditional Medical Practitioners

Out of 28 Amchi surveyed 37.71% were falling in the age group of 20–40 years and 64.29% were above 40 years; 39.28% were illiterate and only 60.72% were literate; 60.72% were having monthly income of rupees less than 3,000 and only 39.28% were having an income of rupees more than 3,000 per month. Most of the Amchi (64.29) were doing practice at home and only 35.71% were doing practice at clinics. Amchi who learned the art of healing from their parents were 53.57% while as 46.43% had obtained training from a recognized Amchi training college (Table 8.3).

Perception of Disease in Rural Ladakh

The study pertaining to people's perception of disease reveals the level of understanding about a disease. It is evident from the Table 8.3 that about 32.50% of the households surveyed considered the cause of disease due to super natural forces

Table 8.3 Socio-economic characteristics of traditional medical practitioners in Ladakh

Physiographic divisions	No. of Amchis surveyed	Socio-economic characteristics of Traditional Medical Practitioners with %age to total									
		Age (years)		Qualification		Income Rs/month		Place of practice		Source of knowledge	
		20–40	Above 40	Illiterate	Literate	<3,000	>3,000	Home	Clinic	Learned from parents	Trained
Nobra	4	1 (25.00)	3 (75.00)	1 (25.00)	3 (75.00)	3 (75.00)	1 (25.00)	3 (75.00)	1 (25.00)	2 (50.00)	2 (50.00)
Leh	7	3 (42.85)	4 (57.15)	1 (14.28)	6 (85.72)	3 (42.85)	4 (57.15)	2 (28.56)	5 (71.44)	2 (28.56)	5 (71.44)
Changthong	7	2 (28.56)	5 (71.44)	5 (71.44)	2 (28.56)	6 (82.72)	1 (14.28)	6 (85.72)	1 (14.28)	6 (85.72)	1 (14.28)
Total for Leh District	**18**	**6 (33.33)**	**12 (66.66)**	**7 (38.89)**	**11 (61.11)**	**12 (66.66)**	**6 (33.33)**	**11 (61.11)**	**7 (38.89)**	**10 (55.55)**	**8 (44.45)**
Zanskar	5	2 (40.00)	3 (60.00)	2 (40.00)	3 (60.00)	3 (60.00)	2 (40.00)	4 (80.00)	1 (20.00)	4 (80.00)	1 (20.00)
Suru	3	1 (33.33)	2 (66.66)	2 (66.66)	1 (33.33)	1 (33.33)	2 (66.66)	1 (33.33)	2 (66.66)	1 (33.33)	2 (66.66)
Drass	2	1 (50.00)	1 (50.00)	Nil	2 (100.00)	1 (50.00)	1 (50.00)	2 (100.00)	Nil	-	2 (100.00)
Total for Kargil District	**10**	**4 (40.00)**	**6 (60.00)**	**4 (40.00)**	**6 (60.00)**	**5 (50.00)**	**5 (50.00)**	**7 (70.00)**	**3 (30.00)**	**5 (50.00)**	**5 (50.00)**
Total for Ladakh	**28**	**10 (37.71)**	**18 (64.29)**	**11 (39.28)**	**17 (60.72)**	**17 (60.72)**	**11 (39.28)**	**18 (64.29)**	**10 (35.71)**	**15 (53.57)**	**13 (46.43)**

Source: Based on data obtained from field survey – 2015

Page 68 G. M. Rather

Table 8.4 Perception of disease by sample households in Ladakh

Physiographic divisions	No. of households Surveyed	Perception of disease (No. of households with %age to total				
		Disease due to nature	Disease due to Bad luck	Disease due to Black magic	Disease due to change in food	Disease due to bad air
Nobra	40	12 (30.00)	8 (20.00)	6 (15.00)	10 (25.00)	4 (10.00)
Leh	52	16 (30.77)	8 (15.38)	6 (11.54)	10 (19.23)	12 (23.08)
Chag thong	28	8 (28.57)	4 (14.33)	10 (35.71)	4 (14.28)	2 (7.14)
Total for District Leh	**120**	**36 (30.00)**	**20 (16.67)**	**22 (18.33)**	**24 (20.00)**	**18 (9.00)**
Zanskar	24	8 (33.34)	4 (16.67)	10 (41.66)	2 (8.33)	--
Suru	16	8 (50.00)	4 (25.00)	4 (25.00)	--	--
Dras	40	12 (30.00)	8 (20.00)	6 (15.00)	10 (25.00)	4 (10.00)
Total for District Kargil	**80**	**28 (35.00)**	**16 (20.00)**	**20 (25.00)**	**12 (15.00)**	**4 (5.00)**
Total for Ladakh	200	64 (32.00)	36 (18.00)	42 (21.00)	36 (18.00)	22 (11.00)

Source: Based on data obtained from field work - 2015

and did not take any medicine. They believed that a curse of God caused disease and thus prayed to Almighty for recovery. Perception of disease due to black magic was noted in 22.50%, Perception of disease due to change in food and perception of disease due to bad air was noted in 16.55% and 11.25% of sample households respectively. A considerable variation in different types of perception of disease has been identified in the sample villages (Table 8.4).

Management of Health Problems in Rural Ladakh

Inter-regional differences within the cold desert Ladakh show different types of management of diseases. Analysis of the Table 8.5 shows that the largest percentage of the households surveyed (26.67%) does not report to the hospital or any other type of medical practitioner for a couple of days but practices home remedies. About 16.39% of the households were found discussing the health problem within the family and 12.22% discussed health problems with the friends. A large group of households (24.72%) consulted the traditional practitioners for various health problems. Allopathic medical facilities were attended by only 12.50% of the households and 10% purchased medicines directly from the medical shops.

Discussion

No doubt both Amchi and allopathic medical systems are utilized in Ladakh, may be because of different attitude of people toward different system of medicines as man has been doing same since past but the trend of both, utilization as well as preference was towards Amchi medical system in all the rural areas of Ladakh and the reason could be for centuries the Amchi, the only medical system was accessible to people and is still an important component of public health. However, utilization of Amchi medical system is more in District Leh than that of District Kargil (Table 8.1) and the reason may be socio- economic and cultural differences. Analysis of the data reveals a very good regional contrast in the utilization of medical systems. Amchi medical system is more dominant in the regions of Changthong and Nobra of Leh district and Zanskar region of Kargil district. The preference of Amchi medical system than allopathic medical system more than 50% not only in both the districts but also in all the regions has been noted and the probable reason for higher preference could be no side effects and low cost in Amchi medical therapy. No doubt majority of households were having source of knowledge for utilization of Amchi medical system as a traditional one (56.52%), however there are very good regional differences in the source of knowledge for utilization of Amchi medical system. The reason could be the regional variation in the increasing role of allopathic medical system and professional Tibetan medical system in Ladakh. The results derived from regression models representing relationship between utilization of Amchi medical system and socio-economic variables show very good results. This can be attributed to the fact that socio-economic and cultural aspects

70 G. M. Rather

Table 8.5 Management of health problems by sample households in Ladakh

Health problems	No. of households surveyed	Discuss in the family for home remedies	Discuss with friends	Purchase medicine directly from shops	Consult traditional practitioner	Go to hospital	Don't go for few days
Fever/typhoid	200	45 (22.50)	20 (10.00)	15 (7.50)	30 (15.00)	30 (15.00)	60 (30.00)
Cough/cold	200	25 (12.50)	15 (7.50)	10 (5.00)	20 (10.00)	15 (7.50)	115 (57.50)
Anemia	200	30 (15.00)	35 (17.50)	15 (7.50)	50 (25.00)	20 (10.00)	50 (25.00)
Diarrhea	200	20 (10.00)	50 (25.00)	15 (7.50)	60 (30.00)	40 (20.00)	15 (7.50)
Eye diseases	200	15 (7.50)	20 (10.00)	25 (12.50)	75 (37.50)	25 (12.50)	40 (20.00)
T. B./asthma	200	30 (15.00)	15 (7.50)	10 (5.00)	50 (25.00)	20 (10.00)	75 (37.50)
Stomach pain	200	35 (17.00)	20 (10.00)	20 (10.00)	40 (20.00)	50 (25.00)	35 (17.50)
Fracture	200	50 (25.00)	20 (10.00)	10 (5.00)	70 (35.00)	15 (7.50)	35 (17.50)
Others	200	45 (22.50)	25 (12.50)	15 (7.50)	50 (25.00)	10 (5.00)	55 (27.50)
Total	1800	295 (16.39)	220 (12.22)	135 (7.50)	445 (24.72)	225 (12.50)	480 (26.67)

Source: Based on data obtained from field work -2015

play an important role in the utilization and preference of different medical systems as observed in the cold desert of Ladakh and above all Amchi medical system being embedded in Buddhism very common religion in Ladakh. It can be visualized from the Table 8.2 that the values of regression equation, coefficient of determination (r2) and coefficient of correlation are very high for the socio-economic variable age, followed by level of income and level of education. Average rate of change denoted by slope of the regression line is also very large for all variables which may be attributed to the fact of strong faith of the people in traditional medical system.

Persons practicing traditional medical system are called traditional healers and the traditional healers in Ladakh are known as Amchi and the present generation of Amchi is the 6th generation of unbroken family. Socio-economic characteristics of Amchi in different physiographic divisions of Ladakh depict some interesting results. All of them were male with a majority falling above the age of 40 years. A large percentage of Amchi were illiterate and have learned art of healing from their parents because it is their traditional medical system and its practice is handed over from one generation to another. To become an Am chi is not easy but it takes some years to become an expert Amchi and the new Amchi is first examined by a group of expert Amchi in front of villagers in a ceremony for registration. A very good percentage of Am chi were formally trained as Amchi like monks used to proceed Lassa to get their training , and even Ladakh Society for Tibetan Medicine (LSTM) offers four year diploma course (Dusrapa Program) and six year degree course (Katchupa Program) at Central Institute for Buddhist Studies in Ladakh. Himachal Pradesh and from Tibetan Institute Leh. It is also evident from the table that most of the Am chi practitioners were doing their practice at home except some few were doing practice at clinics and the clinics were located only in urban areas of Leh and Kargil districts. Majority of the Amchi were having income level above Rs. 3,000 per month which is an indication of good response of the patients towards this medical system.

Since the time immemorial the Himalayan flora has been a major source of medicinal plants and Ladakh range is not an exception to it. No doubt the region has a very rich herbal wealth (Bhat and Wanchoo, 2004) because of wide variation of climate, altitude and soil. There is a very good use of local herbs in different diseases but the problem lies in the fact that majority of herbs are harvested from wild which may be the cause of their depletion and even extinction if conservation not taken on priority and our results were not contradictory to that of LSTM that certain species are under threat of extinction (Nautiyal and Purohit, 2000). The region is characterized by its own floristic composition and the man- plant relationship has been, to a great extent shaped by its traditional culture.

Perception of disease among the sample households reveals the traditional way of understanding of the cause of diseases. All the types of perception of disease like disease due to nature, disease due to bad luck, disease due to magic and disease due to bad air and bad food are based on traditional background. Traditional way of perception of disease in all the regions of district Kargil is higher than that of Leh and the reason could be that district Kargil is more backward than that of district Leh.

Management of health problems among the sample households is also a traditional one. Different practices of treatment of diseases like discussion in the family, discuss with friends, buying medicines directly from medical shop, not going for hospital for many disease and going to traditional healer are all based on traditional knowledge.

Conclusion and Suggestions

The present study leads to the conclusion that traditional knowledge plays an important role in the medical therapy of Ladakh. Amchi medical system is dominant both in utilization and preference especially in rural areas of Ladakh and people have a very good traditional knowledge behind it. Traditional healers are known as Amchi and being an Amchi in Ladakh had been a tradition as most of the Amchi have adopted this tradition from their ancestors and only a few of them obtained this knowledge from different trainings. Perception of diseases and management of various diseases is also based on traditional knowledge. The study reveals that Amchi medical system is well rooted in Ladakh as it is socially, culturally and environmentally close to these people and its development in such a socially and culturally distinct area would strength health care sector. The region has a rich herbal wealth but no attention is paid by the Government for its preservation and there is every chance of its extinction. For the better health care in Ladakh, the following suggestions are recommended;

- Amchi medical system should have clinical efficacy, even if it is practiced by a patient who had no inclination to any form of religious belief, practice or faith. It needs development that too on professional lines so as to make it more scientific and effective. Training programs should be arranged for Am chi by the Government from time to time.
- LSTM should be provided full support for the development of Amchi medical system in Ladakh.
- There is a bad need of integration of allopathic and Amchi medical systems both at institutional and man power levels in such tribal and far-flung area. Job opportunity must be provided by the government to the professional Amchi under NRHM Scheme. There should be one post of Am chi Practitioner at each Allopathic Health Care Facility.
- Ladakh region is ecologically viable for the growth of herbs. Provisions should be made for the conservation of ethno -medical plants in the region. The conservation of medicinal plants is vital to the continuity of Amchi medical system.
- Last but not least is the provision for strengthen of Research in Medicinal plants in the region that will bring efficiency in medicines prescribed by the Amchi.

Acknowledgement

The author is highly thankful to Sonam Dawa Lonpo, Chief Medical Officer, Tibetan Medical Institute Leh for providing valuable information. I am also thankful to different Amchi practising Amchi medical system in Ladakh for sharing their knowledge and experience during field survey.

References

Akhtar R and Rather G M, (1999) Health Care Behavior in Dards, *Journal of Central Asian Studies* 10(1), pp.121–130.

Bhardwaj S M, (1985) Attitude of people towards different systems of medicine, A survey of four villages In Punjab, In *Geographical Aspects of Health and Disease in India*, edited by Akhtar R and Learmonth A T A, Concept Publishing Company, New Delhi, pp.337–359.

Bhat G M and Wanchoo I A, (2004) High altitude medicinal plants and Amchi system of medicine in Ladakh, In *Geography of Health, A Study in Medical Geography*, edited by Izhar N, Ashish Publishing House, New Delhi, pp.249–258.

Chandri B, (2002) Traditional Knowledge and Wisdom- Relevance in the Present Day World, *Geographical Review of India*, 64(2), pp.157–164.

Choedak T, (2000) The Mystic Heart of Healing, *Newsletter*, Tibetan Medical and Astrological Institute of Dalai Lama, *Dhramshala*, 8(2), pp.10–11.

Fonaroof L S and Fonaroff A, (1985) Cultural environment of medical geography in rural hindu India, In *Geographical Aspects of Health and Disease in India*, edited by Akhtar R & Learmonth A T A, Concept Pub. Company, New Delhi, pp.315–335.

Gesler W M, (1991) Cultural aspects in health care geography, In *Environment and Health, Themes in Medical Geography,* edited by R Akhtar, Ashish Publishing House, New Delhi, pp.511–521.

Good C M, (1977) Traditional medicine: An agenda for medical geography, *Social Science and Medicine,* 11, pp.705–713.

Hunter J M and Akhtar R, (1991) The challenge of medical geography, In Environment and Health: *Themes in Medical Geography*, edited by Akhtar R, Ashish Pub. New, Delhi, p.32.

Husain M, (1998) *Geography of Jammu and Kashmir*, Rajesh Publications, New Delhi, p.12.

Hyma B, Ramesh A and Srivastava N, (1986) *Indigenous Practice of Sidah Medicine in South India,* manuscript.

Nautiyal M C and Purohit A N, (2000) Cultivation of Himalayan aconites under polyhouse conditions, *Current Science*, 81(5), pp.1062–1063.

Paul B K, (1985) Approaches to medical geography: A historical perspective, *Social Science and Medicine*, 17, pp.399–405.

Phillips D R, (1984) Healthcare planning in Hong Kong: Can traditional medicine be incorporated, *Proceedings of International Congress on Traditional Asian Medicine,* Surabaya Indonesia.

Ramesh A & Hyma B, (1985) Traditional medical system in practice in an Indian Metropolitan City, In *Geographical Aspects of Health and Disease in India*, edited by Akhtar R & Learmonth A T A, Ashish Publishing House, New Delhi, pp.361–390.

Ramesh A and Hyma B, (1981) Traditional Indian system of medicine as a field of study for medical geographers, *Geographia Medica,* 11, pp.116–140.

Sagwal S S, (1991) *Ladakh- Ecology and Environment*, Ashish Publishing House, New Delhi, pp.9–10.

Stock R, (1980) Health and Health Care in Hausland, *Canadian Studies in Medical Geography*, Atkinson College, Toronto, Ontario, Canada.

Stock R, (1981) Traditional Healers in Rural Hausland, *Geo-Journal*, 5(4), pp.363–368.

Warren D M, (1982), Ghanaian National Policy towards Indigenous Healers, *Social Science and Medicine*, 16, pp.1873–1881.

9 Environmental Compulsions of Pesticide Application in Kashmir and its Impacts on Orchardists

Ishtiaq Ahmad Mayer, Bashir A. Lone,
Manzoor A. Wani and Sheraz A. Lone

Introduction

Exposures to pesticides both occupationally and environmentally cause a wide range of human health problems. The world-wide deaths and chronic diseases due to pesticide poisoning is on increase, it was about 1 million per year during 1990s (Enviro news Forum, 1999), with about three fourths of these occurring in developing countries (Horrigan et al., 2002). At present, India is the second largest producer of pesticides in Asia after China and ranks 12th globally (Mathur, 1999). There has been a steady growth in the production of pesticides in India, from 5,000 metric tons in 1958 to 102,240 metric tons in 1998. In 1996–97, the demand for pesticides in terms of value was estimated to be around Rs. 22 billion (USD 0.5 billion), which is about 2% of the total world market. The pattern of pesticide usage in India is different from that for the world in general. In India out of the total consumption of pesticides, 80% are in the form of insecticides, 15% are herbicides, 1.46% is fungicides and less than 3% are others (Mathur, 1999). The main use of pesticides in India is for cotton crops (45%), followed by paddy, wheat and in horticulture sector. In comparison, the worldwide consumption of insecticides are 44%, herbicides are 47.5%, insecticides are 29.5% and fungicides, 17.5% and others account for 5.5% only.

More than 50% of the population in India (56.7%) is engaged in agriculture and is therefore exposed to the pesticides application used in agriculture (Gupta, 2004). Pesticides used in farms and other orchids are released into the environment and thereby come into human contact directly or indirectly. Humans are exposed to pesticides found in their environment and surroundings (soil, water, air, and food) through different routes of exposure such as inhalation, ingestion, and dermal contact. Intake of contaminated food with pesticide residues is documented to result in highest exposure, about 10^3–10^5 times higher than that arising from contaminated drinking water or air, resulting in acute and chronic health problems, ranging from temporary acute effects such as irritation of the eyes and excessive salivation to chronic diseases such as cancer and reproductive and developmental disorders (Yassi et al., 2001; Zhang et al., 2011).

The high-risk groups exposed to pesticides include production workers, formulators, sprayers, mixers, loaders, and agricultural farm workers. During

DOI: 10.4324/9781003329459-9

manufacture and formulation, the possibility of hazards may be higher because the processes involved are not risk free. In industrial settings, workers are at increased risk since they handle various toxic chemicals including pesticides, raw materials, toxic solvents, and inert carriers. Role of pesticides danger was first studied by American scientist **Rachel Carson** who in her book "Silent Spring" mentions the sudden death of birds caused by indiscriminate spraying of pesticides (DDT). Developing countries use only 20% of the world's pesticide but have reported 99% deaths from pesticide poisoning (Kesavachandran et al., 2009), indicating thereby the inappropriate handling of pesticides.

The credits of pesticide use include enhanced economic potential in terms of food production and amelioration of vector-borne diseases. However, poor agricultural practices adopted by the farmers including the extensive usage of pesticides with the thinking "if a little is good, more will be better" and adoption of insufficient waiting periods before harvesting have resulted in widespread environmental contamination. Among the Indian states, Andhra Pradesh, Uttar Pradesh, Punjab, and Jammu & Kashmir are amongst the highest consumers of pesticides. Pesticide use per hectare is highest in Punjab (923 g/ha) against the national average of 570 g/ha (Devi et al., 2017; Sengupta, 2011). Malwa region of Punjab is described as India's "cancer capital" due to abnormally high number of cancer cases, which have increased three fold in last ten years (Mittal et al., 2014). In India, poisoning due to pesticides was first reported in 1958 in Kerala where over more than 100 people died after consuming parathion contaminated wheat flour and the proportion has been quite high in last ten years as well (Karunakaran, 1958). In an incident at Mashrakh village in district Chhapra of Bihar, 25 school children died after eating a meal provided under mid-day meal scheme. Later, the cooking oil used was reported to be contaminated with monochrotophos (Anonymous, 2013). Acute pesticide poisoning symptoms include allergies, hypersensitivity, giddiness, double vision, headache, dermal abrasions, etc. Mexican farm workers who used pesticides mainly as organophosphates, triazines and organochlorine compounds showed acute poisoning (20% of the cases) and diverse alterations of the digestive, neurological, respiratory, circulatory, dermatological, renal, and reproductive system probably associated to pesticide exposure (Chien et al., 2012; Payan et al., 2012; Ejigu et al., 2005). Long-term effects associated with pesticides include leukemia, lymphomas, soft tissue sarcomas, brain, bone, and stomach cancers, damage to the central and peripheral nervous system, reproductive disorders, birth defects, disruption of the immune system, and death (Michael et al., 2015).

Pesticide Use in Kashmir

The agro-climatic conditions of Valley of Kashmir are ideal for fresh and dry fruit production, which is one of the major economic source of the region. The fruit production area spreads over around a 0.2 million hectares of land, of which 0.11 million hectares, more than 50% is under apple production, involving about 40% population of the Kashmir directly as orchard-farmers, chemical sprayers, etc., and indirectly like children playing in and around orchards, residential houses in orchards (Bhat et al., 2010). A huge quantity of pesticides, insecticides and fungicides (chemicals like chlorpyriphos, mancozeb, captan, dimethoate, phosalone, etc.) are being used by the orchard farmers to spray the plants, fruits and the leaves at different stages of growth to avoid the infestations and destruction of the fruits (Rashid et al., 2010). For the last three decades, the farmers have favored and adapted to the newer synthetic but hazardous fungicides and pesticides, never applied before, to enhance the fruit production by replacing the older relatively non-hazardous inorganic Sulphur. The incidence of the pesticide-related morbidities in Kashmir has shown an upward surge in the last ten years especially in the orchard farming districts (Bhat et al., 2010).

In Kashmir Valley, during the past two decades there has been a substantial increase in the use of pesticides in terms of both volume and value. The demand for agrochemicals depends upon the type of crops grown, farmers knowledge about technologies and their profitability and also upon the availability, affordability and ease in accessing the input and output markets. Among different crops grown in Jammu & Kashmir, apple cultivation is highly capital-intensive with pest control alone accounting for more than 54% of variable costs (Baba et al., 2012). The steady increase in apple productivity in the valley during the past three decades was by and large achieved by increasing the use of fertilizers and pesticides. In the apple-growing belt of the valley, chemicals are being used indiscriminately without considering scientific recommendations (Baba et al., 2012). The choice of chemicals/brand preferences are steered by traders and market functionaries. The excessive/indiscriminate use of pesticides not only increases the cost of apple cultivation but also results in many human health problems and environmental contaminations. These problems get accentuated with the use of spurious chemicals and the existence of a chain of functionaries/unlicensed dealers between firms and farmers (Bhat et al., 2008) Fungicides accounted for the highest (71.1%) sale of total pesticide in the state, followed by insecticides (15.4%) and acaricides (7.7%). The plant growth regulators and weedicides constituted only 3.6% and 2.3% of total pesticide market, respectively. The highest selling pesticide by value was Mancozeb (23.2%), followed by Captan (13.3%), and Fenzaquin (7.0%). Weedicides like Paraquat, Butachlor, and 2, 4- D together constituted only 0.3% of total pesticides sale in the state (Baba et al., 2012).

There is dearth of studies related to pesticide usage and its human health implications India in general and Kashmir in particular. A study of the health effects of acute pesticide toxicity among the cotton growers of India, by (Mancini et al., 2005), is a positive step to fill this research gap. Specific studies dealing with the agricultural practices of Indian farmers regarding pesticide use and its health impacts are needed to inform policy decisions to bring about changes in agricultural practices. Therefore, it is in the backdrop of this literature gap, the present study has been carried out upon the orchard farmers of Kashmir Valley to study the various aspects of pesticide usage in agriculture and its impacts on environment and human health.

Aims and Objectives

The present study has been carried out to achieve the following objectives:

1 To identify the health hazards posed by pesticide use practices among the orchard farmers in the apple growing belt of Himalayan valley of Kashmir.
2 To assess the farmers perceptions of pesticide safety labels, pesticide handling and field spraying practices which might expose them to chemical hazards.
3 To recommend mitigating measures to reduce the hazardous impact of pesticides on orchard farmers health.

Study Area

Being a distinct geographical entity, Kashmir Valley in the state of Jammu & Kashmir has been selected for the present study. The Valley of Kashmir extends from $(33^0\ 22' - 34^0\ 43')$ North Latitudes to $(73^0\ 52' - 75^0\ 42')$ East Longitudes, bordered on the north and west by cease-fire line, south-west by Poonch district, south by Doda and Udhampur districts and north-east by Ladakh district. The region occupies a strategic position in India and is surrounded by Afghanistan in the north-west, Pakistan in the west, China and Tibet in the north-east. The region as a whole covers an area of 322,800 Sq. Kilometers, though an area of 83,808 Sq. Kilometers is under the possession of Pakistan and another 41,500 Sq. Kilometers is occupied by China.

The valley of Kashmir includes all land lying within the water divides formed by Pir-Panjal and the Great Himalayan ranges, which encircle the great synclinal trough occupied by the Jhelum river. The valley lies between Pir-Panjal ranges in the south and the main Himalayan ranges in the north. The fruit production area spreads over around 0.2 million hectares of land, of which 0.11 million hectares (>50%) are under apple production, involving about 40% population of the Kashmir directly as orchard-farmers, chemical sprayers, etc., and indirectly like children playing in and around orchards, residential houses in orchards. In the apple-growing belt of the valley, chemicals are being used indiscriminately without considering scientific recommendations. The incidence of the pesticide-related morbidities in Kashmir has shown an upward surge in the last ten years especially in the orchard farming districts.

Materials and Methods

The present study targeted mainly the orchardist community of the different districts in the Kashmir Valley. A cross-sectional study of 1,214 orchard farmers (1,034 men and 180 women) were interviewed in the study area using a schedule and multi-structured questionnaires prepared and pre-tested. Majority of the farmers contacted were selected from areas of high orchard activity, i.e. Baramulla (21.83%), Shopian (17.13%), and Anantnag (14.09%) while as on the other hand, the lowest were from the areas of Srinagar (3.46%) and Budgam (5.35 %) irrespective of age and gender. The number of farmers interviewed at each administrative unit was decided keeping in view the orchard area of the respective district and the type of dominant crops grown (Table 9.1). The primary exposure relative to pesticides was ascertained based on whether farmers were spraying pesticides or not.

Table 9.1 District-wise area under major horticulture crops and the approximate pesticide consumption in Kashmir Valley

Orchard district	Apple	Pear	Apricot	Peach	Plum	Cherry	Citrus	Other fresh fruits	Orchard area in hectares	Total Cropped Area in hectares	Pesticide consumption in tonnes	Pesticide consumption in litres
Srinagar	1,410 (.98)	290 (4.24)	34 (4.25)	103 (11.43)	80 (5.85)	475 (16.93)	–	210 (12.9)	2,602 (31.7)	8,199	329.9	27060.8
Ganderbal	6,965 (4.88)	381 (5.57)	119 (14.89)	169 (18.75)	262 (19.18)	1,078 (38.43)	–	330 (20.3)	9,304 (47.7)	19,505	1179.2	96854.6
Budgam	12,835 (9)	1,536 (22.46)	27 (3.37)	3 (.33)	453 (33.16)	16 (.57)	–	80 (4.9)	14,950 (20.9)	71,407	1895.6	14960.41
Anantnag	16,971 (11.90)	1,088 (15.91)	146 (18.27)	259 (28.74)	220 (16.10)	147 (5.24)	–	273 (16.8)	19,204 (32.6)	58,830	2435.0	199913.6
Kulgam	18,192 (12.76)	932 (13.63)	111 (13.89)	129 (14.31)	155 (11.34)	36 (1.28)	–	238 (14.6)	19,793 (48.5)	40,597	2509.7	18,192 (12.76)
Pulwama	14,143 (9.92)	943 (13.79)	113 (14.14)	76 (8.43)	97 (7.10)	65 (2.31)	–	7 (.43)	15,444 (27.6)	55,870	1958.2	14,143 (9.92)
Shopian	21,607 (15.16)	396 (5.79)	44 (5.50)	25 (2.77)	36 (2.63)	608 (21.67)	–	164 (10.1)	24,073 (87.37)	27,552	2901.1	21,607 (15.16)
Baramulla	25,203 (17.68)	660 (9.65)	120 (15.01)	56 (6.21)	–	211 (7.52)	20 (86.9)	289 (17.8)	26,559 (83.04)	65,678	3367.6	25,203 (17.68)
Bandipora	61,60 (4.32)	133 (1.94)	27 (3.37)	43 (4.77)	17 (1.24)	56 (1.99)	–	11 (.67)	6,447 (22.2)	29,022	817.4	6,160 (4.32)
Kupwara	19,015 (13.34)	478 (6.99)	58 (7.25)	38 (4.21)	46 (3.36)	113 (4.02)	3 (13.0)	20 (1.23)	19,771 (43.2)	45,718	2506.9	19,015 (13.34)
Total	142,501 (90.7)	6,837 (4.3)	799 (0.5)	901 (0.57)	1,366 (0.87)	2,805 (1.7)	23 (0.01)	1,622 (1.03)	156,954 (37.15)	422,378	199006.6	142,501 (90.7)

Source: J&K Statistical Digest, 2015–16.

A structured questionnaire was designed to collect information on commonly used pesticides and practices, risk perception, attitudes to pesticide labels, precautions, the farmer's source of information about pesticides, and signs and symptoms of illness related to pesticide exposure. The questionnaire was designed in English but the interviews were conducted in the vernacular language. Data were collected through a field survey by face-to-face interviews with farmers. Feedback from the pilot-test interviewers and some respondents had also been considered and the questionnaire was accordingly revised to improve its data-generating ability. In the cross sectional survey, details of signs and symptoms were collected as self reported by farmers. Specific instructions and training were given to the interviewers for collecting data regarding the signs and symptoms of farmers. The interviewers contacted only those farmers who were working in agricultural fields. On encountering farmers working in agricultural fields, the farmers were informed about the purpose of the study and the interviewers obtained verbal consent before proceeding with the interviews. After successful collection of primary data, it was checked, coded, and stored for data entry. Data entry were done in Microsoft Excel (version 4) and further statistical analysis were done using SPSS (version 11).

Calculation of Relative risk (RR value)

The relative risk (RR) is the probability that a member of an exposed group will develop a disease relative to the probability that a member of an unexposed group will develop that same disease. In the present study, RR between the two groups of farmers (sprayers and non-sprayers) was calculated by dividing the percentage of one group (sprayer, which is more affected) by the other, non-sprayer, which is less affected. RR = % (disease exposed)/% (disease unexposed).

Results

Spatial Distribution of Farmers in the Study Area

The present study targeted mainly the agricultural community of the different districts in the Kashmir Valley. The primary exposure relative to pesticides was ascertained based on whether farmers were spraying pesticides or not. Among a total of about 1,214 farmers surveyed across the Kashmir Valley, the majority of them was contacted from Baramulla (21.83%), Shopian (17.13%), and Anantnag (14.09%) while as on the other hand, the lowest was from the areas of Srinagar (3.46%) and Budgam (5.35%). Of the 1,214 farmers interviewed, 852 ("sprayers") reported that they sprayed pesticides themselves; however the remaining 362 were involved in non-sprayer's category. Analyzing the spatial distribution of sprayers and non-sprayers across Kashmir Valley, the study showed that the majority of sprayers were from Baramulla (75.85%) followed by Shopian (73.56%), Anantnag (71.35%), and Kupwara (69.57%) while as the lowest were found in Srinagar (35.38%), Kulgam (64.29%), and Budgam (64.62%)

Table 9.2 District-wise distribution of sample population of farmers in the study area

District	Sprayers	%	Non sprayers	%	Total	%
Baramulla	201	75.85	64	24.15	265	21.83
Kupwara	64	69.57	28	30.43	92	7.58
Bandipora	55	67.90	26	32.10	81	6.67
Ganderbal	43	65.15	23	34.85	66	5.44
Srinagar	24	35.38	18	42.86	42	3.46
Budgam	42	64.62	23	35.38	65	5.35
Pulwama	85	67.46	41	32.54	126	10.38
Shopian	153	73.56	55	26.44	208	17.13
Anantnag	122	71.35	49	28.65	171	14.09
Kulgam	63	64.29	35	35.71	98	8.07
Total	852	70.18	362	29.82	1,214	100.00

Source: Compiled from field study, 2016.

districts (Table 9.2). Likewise in the same way, the majority of the non-sprayer's population was found in Srinagar (42.86%), followed by Kulgam (35.71%) and Budgam (35.38%) while as the other hand, the lowest was found in Baramulla (24.15%) and Shopian districts (26.44%).

Age/Gender Wise Pattern of Farmers

The Table 9.3 depicts the spatial distribution of farmers in the Kashmir Valley both in terms of age as well as gender. Out of the total sprayers (852) interviewed, the majority (840) were from men category, found more (310) in the age group of 31–40 years, however the least male sprayers were from >60 age group. In case of females, the highest number of sprayers was found between the age group of 41–50 while as the negligible found in >60 age category. The study also revealed that among the total non-sprayers (362), 190 were males and 172 were females. The maximum (40) male non-sprayers were found in 51–60 age groups, however corresponding to this scenario, the highest (42) female non-sprayers were found between 31 and 40 age groups.

Knowledge of Pesticides and Pesticide Usage Pattern

Most sprayers (34.15%) relied on retail shop dealers as the information source for knowledge regarding the pesticides they used; 26.53% consulted fellow farmers, and only 10.56% considered the horticulture authority as their source of information. A majority (51.76%) of the sprayers had been spraying pesticides for the preceding more than ten years, (38.15%) over a period of five to ten years and 10.09% for less than five years. The famers when asked about the spraying interval of the pesticides, majority (51.76%) were of the opinion that they applied pesticides

Table 9.3 Age/gender-wise pattern of farmers

Age (years)	Sprayers		Non sprayers		Total
	Male	*Female*	*Male*	*Female*	
<20	53	1	20	18	92
21–30	265	2	27	37	331
31–40	310	3	31	42	386
41–50	128	5	37	36	206
51–60	63	1	40	24	128
>60	21	0	35	15	71
Total	840	12	190	172	1,214

Source: Compiled from field study, 2016.

at an interval of 10–15 days and only 10.09% sprayed pesticides with interval of less than five days .The study thus infers that pesticides in the study area are being sprayed at regular intervals with at high spraying density in the low-lying areas and in the high altitudinal area, their density becomes less due to changes in the geophysical and socio-economic environment. The study further shows that most of farmers (45.77%) sprayed pesticides at a frequency of more than ten times in a year while a low amount (19.95%) of farmers sprayed pesticides with frequency of less than five times in a year. Furthermore, 26.41% of the farmers revealed that the frequency of pesticide application depends up on the type of disease spread in the area. The study thus inferred that the frequency of pesticide application shows quite variations across various districts of the Kashmir Valley being quite high along the low-lying areas and low over the high-altitudinal areas especially Kare was due to less prevalence of the different orchid diseases in the concerned regions (Table 9.4).

Pesticide Handling Practices

Handling of pesticide concentrations and application of diluted formulation require the use of appropriate personal protection equipment as a precaution against exposure. This would include the use of gloves, masks, protective clothes, personal hygiene, appropriate footwear, headgear, etc., as indicated on the pesticide labels (Food & Agricultural Organization: FAO & WHO, 2005). Most farmers in our study were not aware of the health hazards of the inappropriate handling of pesticides. They did not take necessary personal protective measures while handling pesticides; 562 (65.96%) reported that they doesn't wear long sleeved shirts/full pants while handling and spraying pesticides, and only 385' (45.19%) use cotton masks as protective cover equipments during pesticide spraying. Use of gloves was reported by 11.15%, while 31.34% of them also used head cover cloths as a means of protective device. After the spraying of pesticides, the famers in the study area where not taking necessary measures to less down the effects of pesticide exposure. Most of the farmers (53.40%) don't go for bathing immediately after pesticide spraying and also don't change clothes right after spraying and thus became susceptible to different forms of illness related to pesticide exposure. The farmers

Table 9.4 Knowledge of pesticides and pesticide usage pattern among farmers

Variables	Total (N = 852)	%
Information source of pesticides for the Farmers		
Consult dealer	291	34.15
Horticulture consultant	90	10.56
Fellow farmers	226	26.53
Own experience	195	22.89
Other	50	5.87
Total	852	100
Duration of pesticide exposure		
<Five years	86	10.09
Five to ten years	325	38.15
>Ten years	441	51.76
Total	852	100
Spraying interval		
<Ten days	86	10.09
10–15 days	325	51.76
>15 days	441	38.15
Total	852	100
Frequency of pesticide application (yearly)		
<Five times	170	19.95
5–10 times	292	34.27
>Ten times	390	45.77
Total	852	100
Spraying interval		
<Ten days	180	21.13
10–15 days	291	34.15
>15 days	156	18.31
Depends up the disease breakout	225	26.41
Total	852	100

Source: Compiled from field study, 2016.

mixed the different pesticides in a vessel with water or they poured them directly into the spraying can and then mixed them in the spraying can itself. Three hundred and ninety five (46.36%) mixed or diluted pesticides using bare hands. The common alternative to this practice was to use a stick which was reported by only 33.69% of the farmers. The study further reveals that the farmers in the study area continue to perform the different food activities instead of taking necessary precautions during the hours of pesticide spraying; most of the farmers (619, 72.65%) smoke and 647 (75.94%) take meals during the application of pesticides in the study area (Table 9.5).

Other Farming Activities during Pesticide Spraying

Among the total farmers, 307 sprayers and 151 non-sprayers reported that other farming activities continued on the farms while pesticides were being sprayed. During the spraying operation, non-sprayers (41.71%), including women (20.17%), continued to work in the same field, which exposed them to pesticides (Table 9.6).

Table 9.5 Pesticide handling practices practiced by orchard farmers

Variable	Total (N = 852)	%
Wears full sleeved shirt/full pant		
Yes	290	34.04
No	562	65.96
Wears a mask		
Yes	385	45.19
No	467	54.81
Wear a head gear		
Yes	267	31.34
No	585	68.66
Wears special shoes		
Yes	234	27.46
No	618	72.54
Wears an eye glasses		
Yes	74	8.69
No	778	91.31
Wears special gloves		
Yes	95	11.15
No	757	88.85
Do you take bath right after spraying?		
Yes	397	46.6
No	455	53.4
Do you change clothes right after spraying?		
Yes	384	45.07
No	468	54.93
Do you have separate clothing for spraying?		
Yes	274	32.16
No	578	67.84
How do you mix pesticides?		
With bare hands	395	46.36
With stick	287	33.69
Shaking the sprayer	170	19.95
Food habits while spraying		
Smoke during application		
Yes	619	72.65
No	233	27.35
Drink during application		
Yes	603	70.77
No	251	29.46
Eat during application		
Yes	647	75.94
No	205	24.06

Source: Compiled from field study, 2016.

The continuation of pesticide spraying and other farming activities concurrently in the field can lead to direct exposure to pesticides as they may still be dispensed in air (Antonella et al., 2001).

Table 9.6 Continuation of other farming activities while pesticide spraying

	Sprayers (n = 852)	Non sprayers (n = 362)	Total (n = 1,214)
Male	305 (35.80)	78 (21.55)	1,034 (85.17)
Female	02 (0.23)	73 (20.17)	180 (14.83)
Total	307 (36.03)	151 (41.71)	1,214 (100.00)

Source: Compiled from field study, 2016.

Signs and Symptoms of Illness

The farmers in the study area experienced a variety of signs and symptoms related to pesticide exposures. The prevalence of these signs and symptoms was higher among the sprayers as compared to the non-sprayers. Signs and symptoms with high prevalence were excessive sweating (38.06%), burning eyes/stinging eyes/itching eyes (36.49%), fatigue (30.31%), dry/sore throat (27.59%), dizziness (27.18%), and numbness/ muscle weakness/ muscle cramps (24.14%). Among the non-sprayers, signs and symptoms which show high prevalence were fatigue (39.23%), dizziness (33.44%), excessive sweating (33.43%), Burning eyes (31.77%), and blurred vision (30.39%). During the spraying operation, non-sprayers (41.71%), including women (20.17%), continued to work in the same field, which exposed them to pesticides. The higher prevalence of some signs and symptoms among non-sprayers could have been due to their direct exposures to pesticides or due to previous exposures to pesticides on the onset of continuation of other activities while applying pesticides. Awareness to use personal protective measures while handling pesticides is needed among farmers across various districts in the Kashmir Valley. Farmers needs to be encouraged to reduce, if not eliminate, the use of pesticides, with the introduction of incentives to help them shift from synthetic pesticides to bio-pesticides and organic farming (Table 9.7).

Signs and Symptoms and Relative Risk

The analysis of the Table 9.8 highlights the prevalence of signs and symptoms and RR among sprayers and non-sprayers. Among men, there was a higher frequency of signs and symptoms among the sprayers than the non-sprayers, and the exposure factor of spraying pesticides was significantly associated with excessive sweating, burning/stinging/itching eyes and dry/sore throat, with RR values above 1 (1.44, 1.53, and 1.85, respectively). The other signs and symptoms with RR values above 1 were fatigue (1.02), dizziness (1.13), skin redness/white patches on skin/skin scaling (1.47), runny/burning nose (1.64), shortness of breath/cough (1.04), excessive salivation (1.49), nausea/vomiting (1.32), and wheezing (1.27).The signs and symptoms with RR below 1 were Numbness/Muscle weakness/Muscle cramps (0.90), Blurred vision (0.56), Chest pain/Burning feeling (0.93), Tremors (0.85), Stomach Pain/Cramps/Diarrhoea (0.62). Some 20.17% of the women in our study reported that they continued to work while pesticides were being sprayed and thus

Table 9.7 Signs and symptoms of illness among study population of farmers

Health hazard risk	Sprayers (n = 852)	Non sprayers (n = 362)	Total (n = 1,214)
Excessive sweating	40.02	33.43	38.06
Burning/stinging/ itching eyes	38.50	31.77	36.49
Dry/sore throat	29.23	23.76	27.59
Fatigue	26.53	39.23	30.31
Dizziness	24.53	33.44	27.18
Skin redness/white patches on skin/skin scaling	21.83	17.68	20.59
Numbness/muscle weakness/muscle cramps	22.07	29.01	24.14
Running/burning nose	17.96	13.26	16.56
Blurred vision	16.55	30.39	20.68
Chest pain/burning feeling	18.08	27.90	21.00
Shortness of breath/ cough	15.02	23.20	17.46
Excessive salivation	14.67	12.43	14.00
Tremors	13.62	17.40	14.74
Nausea/vomiting	7.51	11.60	8.73
Stomach pain/cramps/ diarrhoea	6.22	15.19	8.90
Wheezing	5.52	11.05	7.17

Source: Compiled from field study, 2016.

it is clear that the women non-sprayers were subjected to different forms of illness related to pesticides more as compared to male non-sprayers. Such exposure could cause a variety of reproductive health problems in women of reproductive age.

Discussion

The present study documented the serious consequences of the indiscriminate use of pesticides for the health of orchard farmers in Kashmir. The study aimed primarily to raise farmer's awareness of the seriousness of the pesticide poisoning occurring in the study area. Most farmers in our study were not aware of the health hazards of the inappropriate handling of pesticides. The use of cotton cloth as protective clothing was common among them. Studies show that wet cotton clothing and cotton masks in fact increase the person's personal absorption rate of pesticides (Kishi et al., 1995). The practice of chewing or smoking, eating, or drinking while spraying is also hazardous to health. This may also indicate that the farmers were symptomatic enough to self-medicate during a pesticide- spraying session, but many are unwilling to follow the necessary precautions, attributing their reluctance to non-availability and high cost of personal protection products, and the prevailing

Table 9.8 Prevalence of signs and symptoms and relative risk among sprayers and non-sprayers

Health hazard risk	Sprayers (n = 840)	Non-sprayers (n = 190)	Relative risk
Excessive sweating	39.40	27.37	1.44
Burning/stinging/itching eyes	37.86	24.74	1.53
Dry/sore throat	30.12	16.32	1.85
Fatigue	26.31	25.79	1.02
Dizziness	24.29	21.58	1.13
Skin redness/white patches on skin/skin scaling	21.67	14.74	1.47
Numbness/muscle weakness/ muscle cramps	22.14	24.74	0.90
Running/burning nose	18.10	11.05	1.64
Blurred vision	16.07	28.95	0.56
Chest pain/burning feeling	17.62	18.95	0.93
Shortness of breath/cough	14.76	14.21	1.04
Excessive salivation	14.88	10.00	1.49
Tremors	13.45	15.79	0.85
Nausea/vomiting	7.62	5.79	1.32
Stomach pain/cramps/diarrhoea	5.83	9.47	0.62
Wheezing	5.36	4.21	1.27

Source: Compiled from field study, 2016.

hot and humid weather conditions. These reasons were similar to those reported from other developing countries such as Indonesia (Antonella et al., 2001).

Combining two or more pesticides, many of which are duplicates (different trade names but the same common name and thus the same active ingredient) should be discouraged. This could result in a dangerous concoction, because mixing of pesticides can alter their chemical properties, thereby increasing their detrimental effects. The combination of use of hazardous pesticides and the absence of appropriate precautions is detrimental to farmers' health (Bolognesi 2011; Salameh et al., 2004). Farmers sometimes returns to the fields for work less than 24 hours after the application of pesticides. The continuation of pesticide spraying and other farming activities concurrently in the field can lead to direct exposure to pesticides, as they may still be dispersed in air (Catano et al. 2006; Antonella et al., 2001). Some 20.17% of the women in our study reported that they continued to work while pesticides were being sprayed. Such exposure could cause a variety of reproductive health problems in women of reproductive age (Bretveld, 2006). This unexpected though direct exposure to pesticides due to women's proximity to the source of exposure needs to be studied further. This aspect of women's being prone to various avenues of exposure has been highlighted in the study done among the cotton growers of India by Mancini et al. (2005).

The number of farmers spraying pesticides for more than a decade was 441 (51.76%), suggesting that many farmers are exposed to pesticides over long durations. This may have chronic health impacts. Young people seem to be engaged in pesticide spraying more than the older people, which may be due to possible attrition of the elderly workforce. Higher-than-normal prevalence of reduced vision (20.68%) among these farmers could be associated with prolonged exposures to pesticides in the present study. The present findings are in accordance with the previous literature by Betarbet et al. (2007) and Solomon et al. (2000) that the farmers in this study experienced a variety of signs and symptoms related to pesticide exposures. Among populations, the prevalence of signs and symptoms related to pesticide exposure was higher among the farmers involved in spraying. The higher percentage of some signs and symptoms among the non-sprayers could be due to their direct exposure to pesticide or due to previous exposure to pesticides. In contrast, the higher prevalence of some signs and symptoms among non-sprayers could have been due to their direct exposures to pesticides or due to previous exposures to pesticides.

Conclusion and Suggestions

The study revealed that most of the orchard farmers in the study area were exposed to hazardous pesticides, did not follow instructions while spraying and exhibited some unsafe practices while using the pesticides. Mostly the orchard farmers in the study area had a low level of knowledge regarding pesticide use. In particular, the farm workers seemed to be unaware of real pesticide risks and they lacked safety education. In addition, the farm workers did not take enough protection measures, which may have exposed them to higher intoxication risks. Although some farm workers were aware of the possible harmful effects of pesticides, as well as by-laws governing their use, but they did not translate this awareness into practice. Awareness to use personal protective measures while handling pesticides is needed among farmers in Kashmir. Farmers needs to be encouraged to reduce, if not eliminate, the use of pesticides, with the introduction of incentives to help them shift from synthetic pesticides to bio-pesticides and organic farming and, also, that there should be stricter enforcement of existing pesticide regulation and monitoring policies to minimize the threats that the farmers' current practices pose to their health and to the environment. Local suppliers are the major distributors of pesticides to the farmers. However, they lack training on usage and storage of pesticides at the shop level, information on pesticide safe handling practices and correct advice to farmers. Regulatory and adequate monitoring policies that can provide adequate extension and advisory services to pesticide distributors on the range of pesticide products available and their uses and handling are recommended. This may improve the quality of pesticide and customer services that are available to the farmers in the community. Government should intensify efforts aimed at registering and controlling distribution of pesticides and banning hazardous ones. This could be achieved through stricter enforcement of existing regulation and monitoring policies. The study also recommended that pesticide manufacturers should be instructed and compelled to exhibit pesticide instructions and warning labels in the language commonly understood by the farmers and other end users, and also to package products in containers that are not attractive for subsequent re-use according to the International Code of Conduct on the Distribution and Use of Pesticides.

In the extreme hot weather conditions, the use of protective gear does not seem to a viable solution to eliminate occupational risks. Educating farmers about the pesticide hazard alone will also do not achieve significant results. The solution seems to be in the replacement of pesticides with non or less toxic alternatives. One example of such alternative is the promotion of integrated pest management approach among the orchardists in the apple-growing belt of Kashmir Valley.

References

Anonymous (2013) Contaminated school meal kills 25 school children. 17–07–2013. <http://articles.chicagotribune.com/2013-07-17/news/sns-rt-us-india-chidren-poison 20130717_1_school-mealchildren- mid-day-meal-scheme>. Accessed 20 April 2014.

Antonella F, Iverse B, Tiramani M, Visentin S, Maroni M (2001) Preventing health risks from the use of pesticides in agriculture. Protecting Workers' health series no. 1. International Centre for Pesticide Safety, WHO, Geneva, Switzerland. <http://www.who.int/occupational_health/publications/pesticides/en/index.html>.

Baba SH, Wani M, Zargar BA, Wani SA, Kubrevi SS (2012) Pesticide delivery system in apple growing belt of Kashmir valley. *Agricultural Economics Research Review* 25(2012).

Betarbet R, Sherer TB, MacKenzie G, Garcia-Osuna M, Panov AV, Greenamyre JT (2007) Chronic systemic pesticide exposure reproduces features of Parkinson's disease. *Nature Neuroscience* 3: 1301–1306.

Bhat AR, Kirmani A, Raina TH, Wani MA, Ramzan AU, Sheikh JI, Shafiq A (2008) Analysis of surgical outcome of Brain tumors in Kashmir – a 26 year experience at SKIMS. *Journal of Medicine Science* 11(3): 152–158.

Bhat AR, Wani MA, Kirmani AR, Raina TH (2010) Pesticides and brain cancer linked in orchard farmers of Kashmir. *Indian Journal of Medical and Paediatric Oncology: Official Journal of Indian Society of Medical & Paediatric Oncology* 31(4): 110.

Bolognesi C, Merlo FD (2011) Pesticides: human health effects. *Encyclopedia of Environmental Health* 16: 438–453.

Bretveld RW, Thomas CM, Scheepers PT, Zielhuis GA, Roeleveld N (2006) Pesticide exposure: the hormonal function of the female reproductive system disrupted? *Reproductive Biology and Endocrinology* 4: 30.

Catano C, Singh VK, Mathur N, Rastogi SK et al (2006) Possible mechanism of pesticide toxicity related oxidative stress leading to airway narrowing. *Redox Report* 11: 159–162.

Chien WC et al (2012) Risk and prognostic factors of inpatient mortality associated with unintentional insecticide and herbicide poisonings: a retrospective cohort study. *PLoS One* 7: e45627–e45632.

Devi PI, Thomas J, Raju RK (2017) Pesticide consumption in India: a spatial temporal analysis. *Agricultural Economics Research Review* 3(1). doi: 10.5958/0974-0279.2017.00015.5

Ejigu D, Mekonnen Y (2005) Pesticide use on agricultural fields and health problems in various activities. *East African Medical Journal* 82: 427–432.

Environews Forum (1999) Killer environment. *Environmental Health Perspectives* 107: A62.

FAO/WHO (2005) Food safety risk analysis, part II, case studies. Food and Agricultural Organization of the United Nations (FAO), Rome, pp. 364–379.

Gupta PK (2004) Pesticide exposure-Indian scene. *Toxicology* 198: 83–90.

Horrigan L, Lawrence RS, Walker P (2002) How sustainable agriculture can address the environmental and human health harms of industrial agriculture. *Environmental Health Perspectives* 110: 445–456.

Karunakaran CO (1958) The Kerala food poisoning. *Journal of Indian Medical Association* 31: 204–207.

Kesavachandran CN, Fareed M, Pathak MK, Bihari V, Mathur N and Srivastava AK (2009) Adverse health effects of pesticides in agrarian populations of developing countries (Whitacre DM, ed). *Reviews of Environmental Contamination and Toxicology* 200: 33–51.

Kishi M, Hirschhorn N, Djajadisastra M, Satterlee LN, Strowman S, Dilts R (1995) Relationship of pesticide spraying to signs and symptoms in Indonesian farmers. *Scandinavian Journal of Work, Environment & Health* 21: 124–133.

Mancini F, Van Braggen AHC, Jiggins JLS, Ambatipudi AC, Murphy H (2005) Acute pesticide poisoning among female and male cotton growers in India. *International Journal of Occupational and Environmental Health* 11: 221–232.

Mathur SC (1999) Future of Indian pesticides industry in next millennium. *Pesticide Information* 24(4): 9–23.

Michael CRA, Matthew KR, Matthew RB (2013) Increased cancer burden among pesticide applicators and others due to pesticide exposure. *CA Cancer Journal for Clinicians* 63: 120–142.

Mittal S, Kaur G, Vishwakarma GS (2014) Effects of environmental pesticides on the health of rural communities in the Malwa region of Punjab, India: a review. *Human and Ecological Risk Assessment* 20: 366–387.

Payan-Renteria R, Garibay-Chavez G, Rangel-Ascencio R et al (2012) Effect of chronic pesticide exposure in farm workers of a Mexico community. *Archives of Environmental & Occupational Health* 67: 22–30.

Rashid BA, Wani MA, Kirmani AR, Raina TH, Ramzan AU, Alam S, … Javed S (2010) Malignant brain tumors (brain cancer) in orchard farmers of Kashmir linked to pesticides. *Current Neurobiology* 1(2): 137–150.

Salameh RP, Baldi I, Brochard P, Saleh BA (2004) Pesticides in Lebanon: a knowledge, attitude and practise study. *Environmental Research* 94: 1–6.

Sengupta, N (2011) A train ride to cancer care. *Times of India*, New Delhi, August 16.

Solomon G, Ogunseitan OA, Kirsch J (2000) Pesticides and human health: a resource for health care professionals. Ph*ysicians for Social Responsibility and Californians for Pesticide Reform* 83.

Yassi A, Kjellstrom T, Kok TK, Gudotli TL (2001) *Basic Environmental Health, World Health Organization*. London: Oxford University Press.

Zhang X, Zhao W, Jing R, Wheeler K, Smith GA, Stallones L, Xiang H (2011) Work-related pesticide poisoning among farmers in two villages of southern China: a crosssectional survey. *Public Health* 11: 429.

10 Urbanization and Quality of Human Health in Srinagar City, Jammu and Kashmir

Arshad Ahmad Lone, G. M. Rather and M. Sultan Bhat

Introduction

The human conditions in the urban habitat have degraded throughout the world and the urban environment has become progressively less livable and less attractive to the present urban residents and to possible future migrants to urban areas (Evans, 2002). The urban environment as a habitat of man is a complex and is formed of heterogeneous components. As regard the socio-economic aspect of the quality of urban life, it is the impairment of the organizational and distributive systems by human beings which causes vertical and horizontal inequalities in the access to and provision for the basic needs (World Health Organization, 2000). The result is development of heat islands, increase in the water and air pollution and high population congestion making urban habitat less livable (Jensen et al., 2004). The qualitative degradation of the urban habitat in the developing countries is of a different kind. It basically results from social pollution as a consequence of scarcity, mass poverty and lopsided development (Fakhruddin et al., 2011). A high rate of urbanization with a low level of urban development, disparity between urban population growth and rate of total development and a slow expansion of urban-based industrialization in the background of the mounting pressure on land in rural areas have resulted in the increased stress on urban amenities and services (Datta, 2007). In the conditions of socially stratified society lopsidedness of urban development has accentuated spatial inequalities in the quality of environment and new areas of qualitatively substandard conditions are added (Peng et al., 2000).

Living in a large urban area can impact on our health and sense of wellbeing through access to health services and recreational opportunities. Population growth and economic development put pressure on the sustainability of the natural environment (Goudie, 2013). Environmental pollution, waste generation and management, built-up areas are all important issues to be considered as urban areas grow and develop (Pandey et al., 2012). These are key components to quality of health.

Enhancement of health status of the people is one of the major objectives of the process of development. Urbanization is an important social process underpinning the dynamics of human society, and it is especially impactful in the 21st Century. Generally, urbanization is accompanied by an increase in the proportion of urban to rural population, population growth in built-up areas; with urbanism referring to

DOI: 10.4324/9781003329459-10

the urban lifestyle and its associated social and behaviour features (Gu et al., 2012). Contemporarily, world urbanization has entered a special period with some new features including information cities or smart cities, multi-centred metropolitan areas, and further globalization involving the transmission of novel ideas and risk behaviours beginning in cities (Li et al., 2010).

Urbanization and urban expansion result in urban environmental changes, as well as residents' lifestyle change, which can lead independently and synergistically to human health problems. In particular, uncontrolled urbanization has been associated in some contexts with pollution, social isolation, overcrowding, changes in dietary and physical activity patterns, and inadequate service capacity for providing drinking water, sanitation and waste disposal, all of which raise the risk of harms to population health (Moore et al., 2003).

Sanitation and water supply have a strong effect on living conditions. The World Health Organization estimated that, even 70–80% of the developing countries' hospital beds are occupied by patients with waterborne diseases. Normally efficient water supply is positively linked with higher per capita income and therefore to urban areas. Although, the services in the city can be even worse than in rural areas, especially in slums and poor settlements (HABITAT-II, 1996).

Literature Review

Various scholars have worked and recognized Quality of health as an important construct in a number of social and medical science disciplines. However, each academic field has developed somewhat different approaches to investigate the construct of quality of life. Some notable contributions are as under:

> Smith (1979) in his book Human geography: A welfare approach used the indicators such as health, education, employment, time & leisure, housing conditions, and security to measure the quality of life and identify areas of deprivation in cities. The author also attempt to use data on cities to produce a ranking of the cities in terms of their Quality of Life.

Pacione (2001) in his book urban geography: A global perspective used different indicators to show the difference in housing quality between high- and low-income countries. The indicators which authors used are floor area per person (m^2), persons per room, percentage of dwelling units with water connection, percentage of permanent structures. He concludes that, on average city obscure high levels of overcrowding with in cities particularly in low-income areas.

Alison et al. (2002) provide an accessible, up to date and authoritative overview of the measurement of Quality of Life in health care. It brings together the work of authors from medicine, nursing, ethics and statistics. Besides this, a number of scholars have worked on to measure the Quality of health in various urban centres of the India (David and Philip, 1994; Akhtar, 2002; Galea and Vlahov, 2005; Agarwal et al., 2007; Butsch, 2008; Harpham, 2009; Gong et al., 2012).

Study Area

Srinagar is the largest urban centre in terms of area and population amongst all Himalayan urban centres. Srinagar city is located at an average elevation of 1,580 meters above mean sea level and it is spread over in the heart of the oval shaped Valley of Kashmir. It is situated between 33°59'14" N to 34°12'37" N latitude and 74°41'06" E to 74°57'27" E longitude.

Databases and Methodology

Data Base

The study is based on both primary and secondary sources of data. Since the universe for the study was large, a random sampling method of ten wards in five clusters of the Srinagar city was selected. The sample taken consisted of 300 sample households from different socio-economic background of the Study area. The Secondary data was collected through various official reports: Census of India, Srinagar Municipal Corporation (SMC), Chief Medical Officer Srinagar (CMO), Water Works Division Master Plan Srinagar, Public Health Engineering Srinagar (PHE) and Integrated Disease Surveillance Programme (IDSP) Srinagar.

Delineation of Study Area

The study area was delineated on the Srinagar municipal corporation map by geo-referencing SOI topographic maps on scale 1:50,000 with the help of ERDAS imagine 9.0 and then digitizing it with the help of ArcView 3.2a. The study area was divided into five zones, using the Srinagar Municipal Corporation census data of 2011. Further, the data regarding quality of health of the study area was processed in Arc View 3.2a for the preparation of various thematic maps.

Basis of Clustering

The study area was first grouped into five clusters on the bases of household density for the purpose of sample survey. A sample of ten percent (10%) representing wards has been taken in each cluster, which comprises almost two (2) wards as sample in each cluster. Then two percent (2%) household samples have been taken in each sample wards of the respective clusters as shown in Table 10.1. Average household density was worked out in each cluster, in which one sample ward was taken below average and another sample ward was taken above average.

A household sample survey was conducted with the help of an interview schedule. First, a draft schedule was prepared, keeping all the broad aspects of different variables that have relationship with the quality of health. The household sample survey was conducted in five identified clusters of Srinagar city.

Table 10.1 Sample frame

Clusters	No. of households/km²	No. of wards	Sample wards	Total no. of households	Sample households
Zone I	Less than 500	16	2	4,976	69
Zone II	501–1,000	12	2	3,714	57
Zone III	1,001–1,500	13	2	4,670	66
Zone IV	1,501–2,500	12	2	4,017	55
Zone V	Above 2,500	15	2	3,632	N54
Total		**68**	**10**	**21,009**	**300**

Table 10.2 Variables for quality of health and related facilities

Variables
Z_1 Morbidity
Z_2 Health institutes per lakh population
Z_3 Percentage of houses with regular water supply
Z_4 Percentage of availability of dustbins
Z_5 Percentage of removal of dustbins
Z_6 Percentage of sewerage facility
Z_7 Percentage of closed sewerage
Z_8 Percentage of disposal of wastes in dustbins
Z_9 Percentage of disposal of wastes on road side
Z_{10} Percentage of disposal of wastes on neighboring plot

Data Analysis

The data is analysed and interpreted with the help of following sources, tools and techniques:

I Survey of India Toposheets
II SMC Ward Map Srinagar City
III GIS Software's like Arc View 3.2a and ERDAS imagine 9.0
IV Composite Index Technique

Variables Used to Analyse the Quality of Human Health in Srinagar City

In the present analysis, ten variables have been chosen to capture the scenario of the existing Health in Srinagar city. Table 10.2 gives an idea of the chosen variables. The levels of quality of health thus obtained have been plotted in respective figures.

Results and Discussions

Population Growth

During the last century the population of Srinagar city (1901–2011) has been phenomenal, it increased from 122,618 persons in 1901 to 1,147,613 persons in 2011, indicating nine-fold increases amounting to 692.18% growth with a net increase of 1,024,995 persons. The pattern of decadal growth however, has not been uniform. In the early decades from 1901 to 1961, the growth has been slow due to the low growth rates which has declined from 22.46% in 1931 to 15.71% in 1961. This decline in the growth rate could be attributed to the political unrest and partition of the subcontinent in 1947 which led to the large scale migration of people. It was after 1961 that a new phase of growth of population commenced. The main factors responsible for this accelerated population growth during this period have been in migration, increase in birth rates and fall in death rates. Besides this, the merger of 62 villages in municipal limit in 1971A.D. and the introduction of urban agglomeration concept which brought a number of rural areas under the jurisdiction of Srinagar city are indeed the other factors contributing to the rapid growth of the city population. Subsequently from 1981 to 2011 A.D. the population increased to 971,357 persons in 2001, registering a net growth of 365,355 persons in two decades with a decadal growth rate of 30.14% and 1,147,613 persons in 2011 recording a net addition of 176,256 persons during the last ten years. The ever increasing population coupled with limited space and haphazard growth has resulted in unwieldy expansion of the city which is reflected from the fact that the city size has increased from 12.8 kms^2 in 1901 to 278.1 kms^2 in 2011 therefore, registering an increase of 265.3 kms^2 during the last decade.

The population growth of the city from 1901 to 2011 is shown in Table 10.3.

Table 10.3 Srinagar City: spatio-temporal growth (1901–2011)

Year	Area (km^2)	Population (persons)	Density persons/km^2	Decadal growth rate
1901	12.80	122,618	9,579	
1911	12.85	126,344	9,832	+3.04
1921	14.48	141,735	9,788	+12.18
1931	17.60	173,573	9,862	+22.46
1941	17.60	207,787	11,806	+19.71
1951	29.52	246,522	8,351	+18.64
1961	41.42	285,257	6,884	+15.71
1971	82.88	403,413	4,867	+34.31
1981	208.09	606,002	2,912	+40.13
1991*	243.09	788,680	3,244	+30.14
2001	278.1	971,357	3,492	+30.14
2011	278.1	1,147,613	4,126	+18.15

Source: Srinagar Municipal Corporation.

* Population of 1991 was obtained through interpolation as census was not conducted in J&K.

The Figure 10.1 highlights the relationship between population growth and spatial extent of Srinagar city from 1901 to 2011.

Population Distribution

There is uneven distribution of population in Srinagar city. Some zones are most populated where as some zones are sparsely populated. The maximum population is in Zone I comprising 287,042 persons, while least population is settled in Zone IV.

Population Density

The gross density of population of the study area given in Table 10.4 is 4,126 persons per km^2. However, there are marked spatial variations in the density of population within the city. It is very high in older parts (Zone V) located in the central city comprising the wards: Ali kadal, Chattabal, Barbarshah, Safa kadal, Alochibagh, Khankai moula, Islamyarbal, Zindshah sahab, Ganpatyar, Nawab bazar, Syed Ali Akbar, Aqilmir khanyar, Jamia masjid, Malik angan, SR Gunj. High Density areas (Zone V) which form most congested parts in the city covers an area of 9.4 Km^2 accommodating 226,645 persons having population density of 24,111 persons/Km^2. The density drops radically just outside the older city which includes Zone IV Comprising the following wards: Shaheed gunj, Soura, Sheikh Dawood colony, Kawdara, Nowshera, Idd Ghah, Jogilankar, Chanapora, Jawaha-rnagar, Khawaja bazar, Hassan Abad, and Batamallo. This ward covers 16.9 Km^2 of city accommodating 200,962 persons. It is surrounded by medium density population. This zone (Zone III) includes the following wards: Karannagar, Lalbazar, Bemina (A), Natipora, Parimpora, Mehjornagar, Tarabal, Zadibal, Qamarwari,

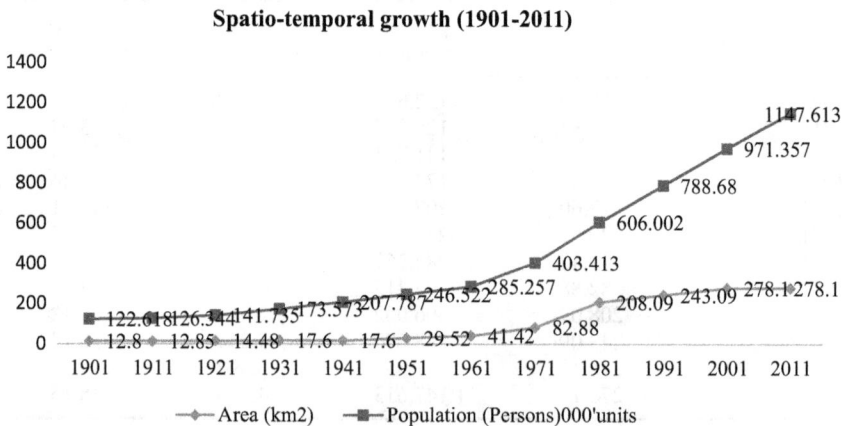

Figure 10.1 Spatio-temporal growth (1901–2011).

Source: Srinagar Municipal Corporation.

Table 10.4 Population density of Srinagar city

Clusters	Area (km²)	Area (%)	Population (persons)	Population density (km²)	Population (%)
Zone I	174.6	62.78	287,042	1644	25.01
Zone II	50.4	18.12	231,016	4583.65	20.13
Zone III	26.8	9.64	201,948	7535.37	17.60
Zone IV	16.9	6.08	200,962	11891.24	17.51
Zone V	9.4	3.38	226,645	24111.17	19.75
Total	278.1	100.00	1,147,613	4126.62	100.00

Source: Compiled from SMC Data, 2011.

Magarmalbagh, Buchpora, Zonimar and Makhdom sahib. These wards cover an area of 26.8 Km² of Srinagar city accommodating 201,948 persons. Zone II which surround this zone have moderately low density of 4,583 persons per Km², cover an area of 50.4 Km² and accommodate 231,016 persons. This zone consist Dalgate, Pandrethan, Hyderpora, Nundresh colony, Ahmad nagar, Rajbag, Bemina (B), Lalchok, Umarcolony, Zainakot, Hazratbal and Madinsahab wards. The lowest densities are found in the outer fringes which are predominantly dominated by rural landscape with intervening village settlements. These areas cover an area of 174.6 Km², accommodating 287,042 persons with a density of 1,644 persons per Km² as shown in Figures 10.2 and 10.3. These outlying suburbs/outer fringes also form the potential areas for future development of city.

It is clear that about 9% area of Srinagar city, i.e. (high density area, Zones IV & V) constitutes about 37% population of Srinagar city, while as 63% area of Srinagar city, i.e. (low density area Zone I) accommodates only 25% population of Srinagar city, and 28% moderate density area (Zones II & III) accommodates 38% of city population as shown in Figures 10.2 and 10.3. Such a scenario of density calls for a policy of rationalization of density pattern in the city including

Area Under clusters

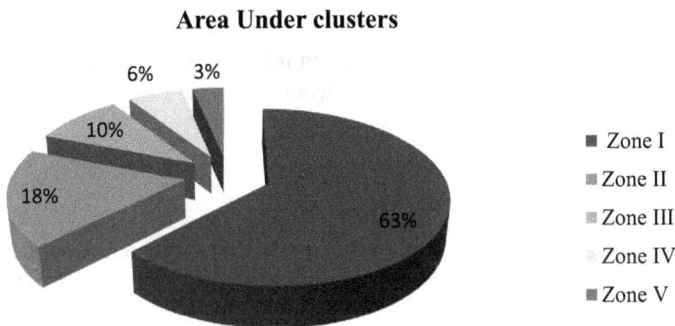

Figure 10.2 Area under clusters.
Source: Based on data obtained from SMC, 2011.

Population Distribution in Clusters

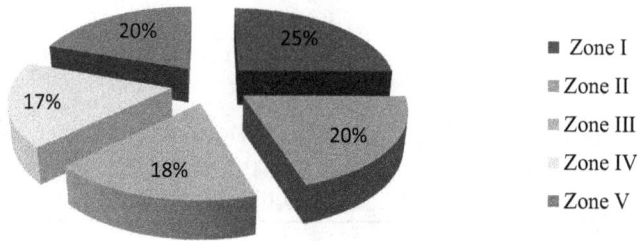

Figure 10.3 Population distribution in clusters.

Source: Based on data obtained from SMC, 2011.

Population Density / km2

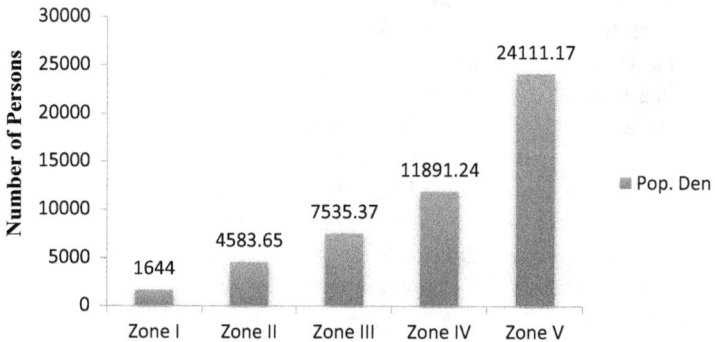

Figure 10.4 Population density/km²

Source: Based on data obtained from SMC, 2011.

redensification of low-density residential areas and decongestion of older parts to make optimal use of land resource and urban infrastructure in the city.

Population density per km² increases from 1,644 persons in Zone I to 24,111 persons in Zone V as highlighted in Figure 10.4.

Household Distribution and Density

There is uneven distribution of households in Srinagar city. Some zones are having densely households where as some zones have sparsely households. The highest number of households is in Zone I comprising 43,690 households followed by Zone II and Zone V having 36,517 & 35,021 households respectively, while least households are settled in Zone IV and Zone III comprising 31,320 and 31,665 households given in Table 10.5.

Household Density

The gross density of households of the study area works out to be 641 per km². However, there are marked spatial variations in the density of household within the city, highlighted in Figure 10.5. It is very high in older parts (Zone V) located in the central city. High-density areas (Zone V) which form most congested parts in the city cover an area of 9.4 Km² accommodating 35,021 households having household density of 3,725 households/Km². The density drops radically just out-side the older city which includes Zone IV. It covers an area 16.9 Km² of city accommodating 31,320 households having density of 1,853 households/km². This zone is surrounding by medium density household. It covers an area of 26.8 Km² of Srinagar city accommodating 31,665 households of density 1,181 households/km². Zone II which surround this zone have moderately low density of 724 households per Km², cover an area of 50.4 Km² and accommodate 36,517 households. The lowest density is found in the outer fringes which are predominantly dominated by rural landscape with intervening village settlements. These areas cover an area of

Table 10.5 Household distribution and density

Clusters	Area (km²)	Households (00' units)	Household density (km²)
Zone I	174.6	436.90	250.23
Zone II	50.4	365.17	724.54
Zone III	26.8	316.65	1181.53
Zone IV	16.9	313.20	1853.25
Zone V	9.4	350.21	3725.64
Total	278.1	1782.13	640.82

Source: Based on data obtained from SMC, 2011.

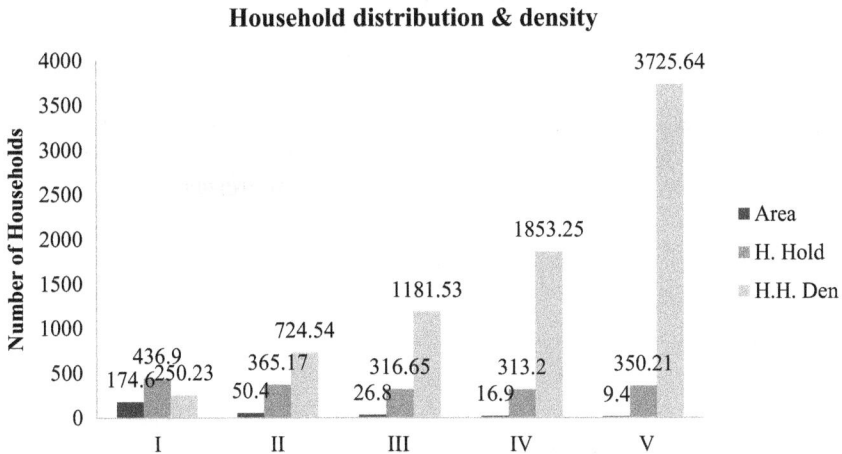

Figure 10.5 Household distribution and density.
Source: Based on data obtained from SMC, 2011.

174.6 Km², accommodating 43,690 households with a density of 250 households per Km². These outlying suburbs/outer fringes also form the potential areas for future development of city.

Density is a controversial and much-used term in planning vocabulary. At its simplest, residential density is the number of units in a given area. There are no agreed upon standard definitions of density; rather each location and profession has come up with an idiosyncratic view of density. Planners and planning documents commonly refer to housing density and population density as low, medium and high densities. However, it is difficult to find a commonly accepted definition for low, medium and high density. Basically, residential density is calculated based on the land area.

Morbidity

In epidemiology and actuarial science, the term "morbidity rate" can refer to either the incidence rate, or the prevalence of a disease or medical condition. Table 10.6 reveals the morbidity of the Srinagar city.

The highest numbers of patients are recorded in Zone V which is 203,971; in Zone II it is 70,296 while as in Zone III it is 130,186 followed by Zones IV and V comprising 123,117 and 203,971 patients in the 2017 as shown in Figure 10.6. While as the maximum morbidity is recorded in the Zones V, III and IV having 0.89, 0.64 and 0.61, and least is recorded in Zones II and I having 0.30 and 0.47 values respectively as indicated in Figure 10.6.

The maximum number of health institutes per lakh population is recorded in Zone I, having 10.45 health institutes followed by Zone II comprising 9.09 health institutes. The least health institutes are recorded in Zone V comprising 1.32 health institutes per lakh population as shown in Tables 10.7 and 10.8.

It is obvious from the Figure 10.7 that the number of health institutes per lakh population decreases from Zone I to Zone V the reason could be the.

Water Supply

Organized water supply to Srinagar city was introduced towards beginning of 20th century, since then there has been gradual and steady augmentation, improvement and requisite extensions to cover expanding outskirts of Srinagar city.

Table 10.6 Morbidity of Srinagar city

Clusters	Population (persons)	No. of patients	Morbidity
Zone I	287,042	136,588	0.47
Zone II	231,016	70,296	0.30
Zone III	201,948	130,186	0.64
Zone IV	200,962	123,117	0.61
Zone V	226,645	203,971	0.89
TOTAL	1,147,613	664,158	0.57

Source: Chief Medical Officer, Srinagar 2017.

No. of Patients

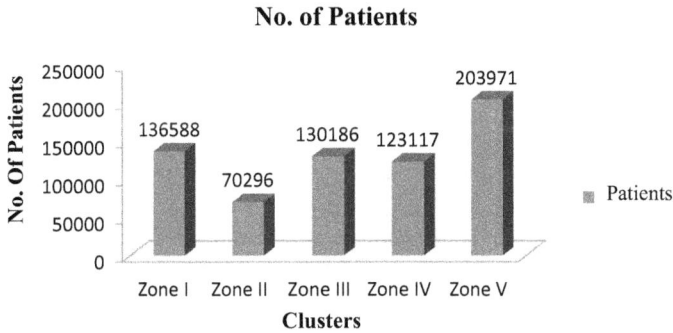

Figure 10.6 No. of patients.

Source: Based on data obtained from Chief Medical Officer 2017.

Table 10.7 Distribution of health care institutes

Clusters	Population (persons)	No. of health institutes	No. of health institutes/ lac population
Zone I	287,042	30	10.45
Zone II	231,016	21	9.09
Zone III	201,948	14	6.93
Zone IV	200,962	11	5.47
Zone V	226,645	3	1.32
Total	1,147,613	79	6.88

Source: Directorate of Health Services Kashmir 2017.

Table 10.8 Sources of drinking water supply

Clusters	Sample wards	No. of sample households	Sources of water supply		
			Tap	Well	River
Zone I	2	69	100	0	0
Zone II	2	57	100	0	0
Zone III	2	66	100	0	0
Zone IV	2	55	100	0	0
Zone V	2	54	100	0	0

Source: Sample Survey, 2017.

Good progress has been made in tap water supply as highlighted in Figure 10.8. From the sample survey, all the households were using the water through tap. None of the respondents were fetching water directly either from well or river.

Though, the entire populations in the city are provided with piped water supply. However, the supply of water is quite in adequate. In many areas of different Zones of city the water is supplied only at fixed interval as revealed in Table 10.9.

Health Institutes per lakh Population

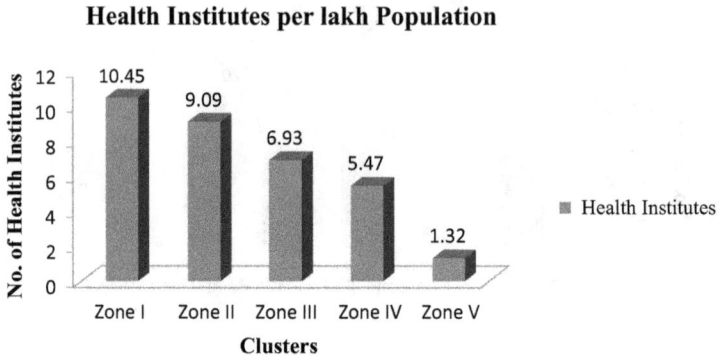

Figure 10.7 Health Institutes per lakh population.

Source: Based on data obtained from Directorate of Health Services Kashmir.

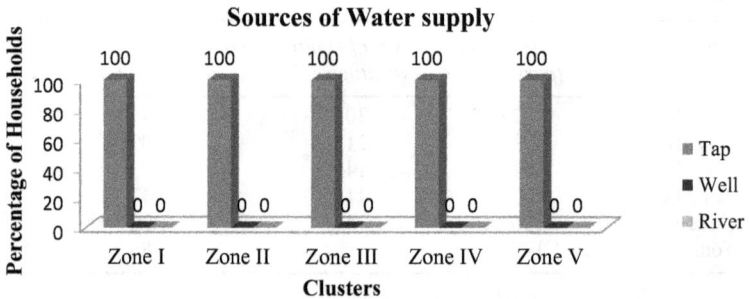

Figure 10.8 Sources of water supply.

Source: Sample Survey, 2017.

Every Zone in the Srinagar city is affected with regular water supply. The highest irregularity is found in Zone V were 87% household gets regular water supply. The cause for the irregularity of water supply in these areas are because of the natural increase In the population and very old pipe lines of water supply schemes, which are not sufficient to cater the demand of present population. Least irregular water supply is found in Zone I, were 97% household gets regular water supply highlighted in Figure 10.9.

Sewerage

Sewage is a water-carried waste, in solution or suspension that is intended to be removed from a community. Also known as wastewater, it is more than 99% water and is characterized by volume or rate of flow, physical condition, chemical constituents and the bacteriological organisms that it contains.

The maximum drainage facility are available in Zones IV, V and II occupying to 98.18, 98.15 and 92.98% households, while as it is least in Zones I and III

Table 10.9 Water supply

Clusters	Sample wards	No. of sample households	Water supply	
			Regular	Irregular
Zone I	2	69	97.10	2.90
Zone II	2	57	94.73	5.26
Zone III	2	66	92.42	7.58
Zone IV	2	55	89.09	10.91
Zone V	2	54	87.04	12.96

Source: Sample Survey, 2017.

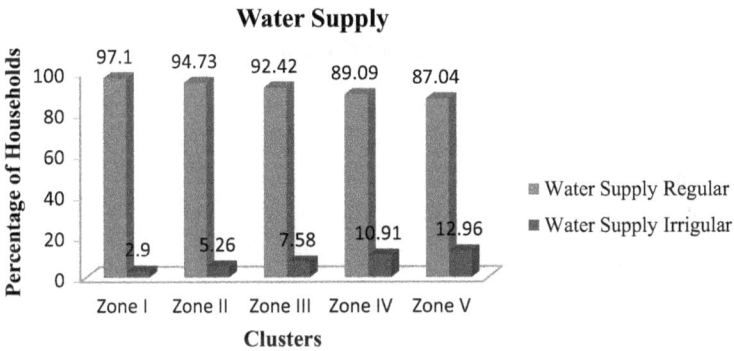

Figure 10.9 Water supply.
Source: Sample Survey, 2017.

occupying 86.96 and 87.88% households respectively. While as on the other side highest number of closed drains are found in the Zones V and II facilitating 94.44 and 92.98% households, followed by Zones IV, III and I facilitating 89.09, 84.85 and 42.03% households in the city respectively shown in Table 10.10.

Though Srinagar city has availability of sewerage facility, but it is not sufficient to drain out sewerage properly highlighted in Figure 10.10. The situation becomes more acute during rainy seasons, because of the narrow, damaged and blockade of the drains.

Sanitation

The availability and disposal of dustbins per day is revealed in Table 10.11.

The maximum dust bins are available in Zones II and III providing facility to 77 and 59% households, while as it is least in Zones V, I and IV providing facility to 24, 26 and 43% households respectively. While as on the other side disposal of dustbins per day are found in the Zones V and IV facilitating 90.94 and 89.09% households, followed by Zones II, III and I facilitating 87.72, 86.36 and 69.57% households in the city respectively as given in Figure 10.11.

Table 10.10 Sewerage facility

Clusters	Sample wards	No. of sample households	Sewerage facility			
			Yes	No	Closed	Open
Zone I	2	69	86.96	13.04	42.03	57.97
Zone II	2	57	92.98	7.02	92.98	7.02
Zone III	2	66	87.88	12.12	84.85	15.15
Zone IV	2	55	98.18	1.82	89.09	10.91
Zone V	2	54	98.15	1.85	94.44	5.56

Source: Sample Survey, 2017.

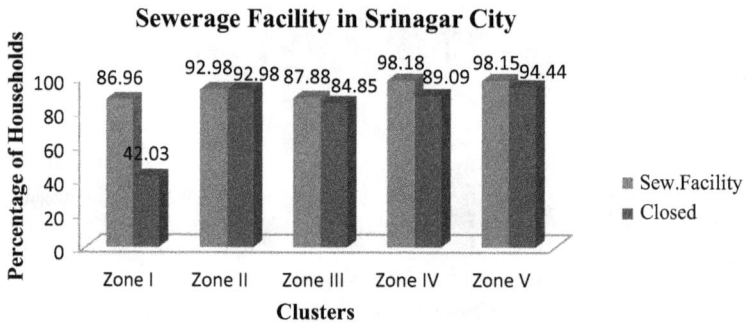

Figure 10.10 Sewerage facility in Srinagar City.
Source: Sample Survey, 2017.

Table 10.11 Availability of dustbins

Clusters	Sample wards	No. of sample households	Availability of dustbins		Disposal of dustbins per day	
			Yes	No	Yes	No
Zone I	2	69	26.09	73.91	69.57	30.43
Zone II	2	57	77.20	22.80	87.72	12.28
Zone III	2	66	59.09	40.90	86.36	13.63
Zone IV	2	55	43.64	56.36	89.09	10.90
Zone V	2	54	24.07	75.93	90.74	9.26

Source: Sample Survey, 2017.

Majority of the population disposes off their solid waste on streets, open space, drains, lanes and storm water drainage. There is inadequate number of collection bins so solid waste gets heaped on road/street side given in Table 10.12. In order to save labour many workers also burn the smaller heaps or dump them into open drains. This system has resulted in unhygienic and poor environmental conditions

Table 10.12 Disposal of wastes

Clusters	Sample wards	No. of sample households	Disposal of wastes		
			Dustbin	Road side	Neig. Plots
Zone I	2	69	14.50	79.71	5.79
Zone II	2	57	59.65	40.35	0
Zone III	2	66	46.97	51.51	1.52
Zone IV	2	55	36.37	63.63	0
Zone V	2	54	22.22	77.78	0

Source: Sample Survey, 2017.

Availability & Disposal of Dustbins

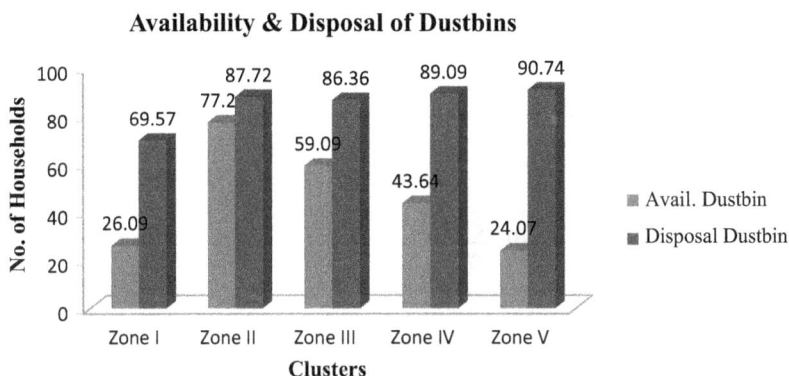

Figure 10.11 Availability & disposal of dustbins.

Source: Sample Survey, 2017.

in the city. Delay in the collection and transportation of waste creates aesthetic problems like smell odour, proliferation of flies and mosquitoes and other vectors resulting in increased disease and morbidity.

The maximum household's disposes off their wastes in dust bins are recorded in Zone II and least in Zone I. The households which dispose off their wastes on road side are highest recorded in Zones I and IV and least in Zone II, while as the highest number of households that dispose off their wastes in neighbour plots are recorded in Zone I shown in Figure 10.12 (Tables 10.13 and 10.14).

The composite index reflects the level of Quality of Health in the Srinagar city, shown in Table 10.11. The table reveals that Zone II has very high-quality in the field of Health facilities with composite index value 18. It is occupied by high government officials like, administrative officers, managers, professors, etc. and big businessmen personals. Houses are concrete and pakka with the green lawns which increases the ventilation, results the high quality of Health. A health institute per lakh population is high as compared to other zones in the city. Availability of dustbins, sewerage facility, etc. in percentage is high which results the very high quality of health as compared to other zones in the Srinagar city.

Disposal of Wastes

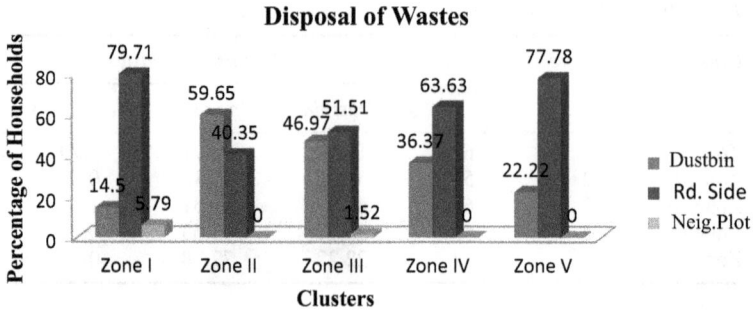

Figure 10.12 Disposal of wastes.

Source: Sample Survey, 2017.

Table 10.13 Selected variables for quality of health

Clusters	Morbidity	Health inst. / lac	Water supply (%)	Avlb. dustbin (%)	Remove dustbin (%)	Sewerage facility (%)	Closed (%)	Dustbin (%)	Road (%)	Neigh. plot (%)
Zone I	0.47	10.45	97.10	26.09	69.57	86.96	42.03	14.50	79.71	5.79
Zone II	0.30	9.09	94.73	77.20	87.72	92.98	92.98	59.65	40.35	0
Zone III	0.64	6.93	92.42	59.09	86.36	87.88	84.85	46.97	51.51	1.52
Zone IV	0.61	5.47	89.09	43.64	89.09	98.18	89.09	36.37	63.63	0
Zone V	0.89	1.32	87.04	24.07	90.74	98.15	94.44	22.22	77.78	0

Source: Compiled from both primary and secondary sources.

Table 10.14 Composite index values for quality of health

Clusters	Z_1	Z_2	Z_3	Z_4	Z_5	Z_6	Z_7	Z_8	Z_9	Z_{10}	Composite index
Zone I	2	1	1	4	5	5	5	5	5	5	38
Zone II	1	2	2	1	3	3	2	1	1	2	18
Zone III	4	3	3	2	4	4	4	2	2	4	32
Zone IV	3	4	4	3	2	1	3	3	3	2	28
Zone V	5	5	5	5	1	2	1	4	4	2	34

Source: Compiled from both primary and secondary sources.

Note: Higher the values lower the Quality of Health and lower the value higher the Quality of Health.

The Zone IV exhibit high level of development in their infrastructures with index value 28. In this zone availability of dustbins, closed sewerage facility is high as compared to other zones which results the second rank in the quality of heath in the Srinagar city.

The Zone III has achieved a medium level of development in the provision of health related infrastructures with index value 32. It is occupied by the medium income groups, removal of wastage per day, sufficient sewerage facility results the medium level of quality of health.

The low level of health status is in Zone V with composite index value 34. It is the core of the city, which lacks all health-related facilities due to substandard environmental sanitation and hygienic condition as a consequence of congestion, overcrowding, low incomes and poor health.

The very low level of quality of Health and its related infrastructures is found in Zone I with composite value 38. It is a mixed population of both rural and urban population. New settlement is in a greater pace that results the lack of urban related amenities like closed sewerage, dustbins, health institutes, etc. in the zone.

Analysis of the data reveals that there is positive correlation $r = + 0.17$ between household density and quality of health, and also positive correlation $r = + 0.57$ between morbidity and quality of health.

Conclusions

The present study analysed and evaluated the quality of health of people in Srinagar city. The study area was clustered into five zones on the basis of household density. The study revealed that there is uneven distribution as well as density of population. Some clusters are most populated while as others are sparsely populated; same is the case in household's distribution and density. The highest numbers of patients are recorded in Zone V and least in Zone II, it is because of the overcrowding of population which results the lack of proper ventilation, lack of proper disposal off sewerage and sanitation. While as the number of health institutes per lakh population decreases from Zone I to Zone V, it is due to the sparse population in Zone I because of leapfrog settlement and high density of population in Zone V as it is the core of the city. All the households in the Srinagar city are using tap water but the supply remains irregular especially in the areas fed on the Doodhganga water supply schemes. Zone V exhibits slum like conditions of in adequate water supply and other sewerage and sanitation facilities are also lacking here. There is also a positive correlation of household density and morbidity with quality of health.

Suggestions

- A health care facility needs to be strengthened for both medical and para-medical staff for improving the quality of health in Zone I.
- Replacement of age-old water supply pipes in the Srinagar city especially in Zones IV and V in a view to cater the present water demand of said zones.
- Construction of drains and Increase in the Municipal dustbins and its related tools and machines to improve the sewerage and sanitation in Srinagar city especially in Zones I and V.

References

(HABITAT-II, 1996).

Agarwal, S., Satyavada, A., Kaushik, S. & Kumar, R. (2007) Urbanisation, urban poverty and health of the urban poor. Status, challenges and the way forward, *Demography India*, 35, 121–134.

Akhtar, R. (2002) *Urban Health in the Third World*, APH. Publishing Corporation, New Delhi.

Alison, C., Irene, H. & Peter R. (2002) *Quality of Life*, Wiley – Blackwell.

Butsch, C. (2008) Access to healthcare in the fragmented setting of India's fast growing agglomerations – a case study of Pune, in: Bohle, H.G. & K. Warner (eds). *Megacities. Resilience and Social Vulnerability*. UNU Institute for Environment and Human Security (UNU-EHS), Bonn, pp. 62–72.

Datta, P. (2007) *Urbanisation in India*, Indian Statistical Institute, Kolkata.

David, R. & Phillips (1994) *Health and Development*, Routledge Publication, London and New York.

Evans, P. (2002) Looking for agents of urban livability in a globalized political economy, in: P. Evans (ed). *Livable Cities? Urban Struggles for Livelihood and Sustainability*. University of California Press, Oakland, CA, USA.

Fakhruddin, S. A. & Khan, M. F. (2011) *Quality of Urban Environment: Some Theoretical and Methodological Considerations. National Geographical Journal of India*, 54(4), 73–82.

Galea, S., & Vlahov, D. (2005). Urban health: Evidence, challenges, and directions. *Annual Review of Public Health, 26*(1), 341–365.

Giles-Corti, B., Vernez-Moudon, A., Reis, R., Turrell, G., Dannenberg, A. L., Badland, H. & Owen, N. (2016) City planning and population health: a global challenge. *The Lancet, 388*(10062), 10–16 December, 2912–2924.

Gong, P., Liang, S., Carlton, E J., Jiang, Q., Wu, J., Wang, L., et al. (2012) Urbanisation and health in China. *Lancet, 379*(9818), 3–9 March, 843–852.

Goudie, A. S. (2013) *The Human Impact on the Natural Environment: Past, Present, and Future*. John Wiley & Sons.

Gu, C. L., Wu, L. Y. & Cook, L. (2012) Progress in research on Chinese urbanization. *Frontiers of Architecture and Civil Engineering in China, 1*(2), 101–149.

Harpham, T. (2009) Urban health in developing countries. What do we know and where do we go?, *Health & Place, 15*(1), March, 107–111.

Jensen, R., Gatrell, J., Boulton, J., Harper, B. (2004). Using Remote Sensing and Geographic Information Systems to Study Urban Quality of Life and Urban Forest Amenities. Ecology and Society *9*(5): 5. [online] URL: http://www.ecologyandsociety.org/vol9/iss5/art5/ La

Li, X. H., Gao, L. L., Dai, L., Zhang, G. Q., Zhuang, X. S., Wang, W., et al. (2010) Understanding the relationship among urbanisation, climate change and human health: a case study in Xiamen. *International Journal of Sustainable Development & World, 17*(4), 304–310.

Moore, M., Gould, P. & Keary, B S. (2003) Global urbanization and impact on health. *International Journal of Hygiene and Environmental Health, 206*(4–5), 269–278.

Pacione, M. (2001) *Urban Geography: A Global perspective,* Routledge Publishers, London and New York.

Pandey, P. C., Sharma, L. K. & Nathawat, M. S. (2012) Geospatial strategy for sustainable management of municipal solid waste for growing urban environment. *Environmental Monitoring and Assessment, 184*(4), 2419–2431.

Peng, X., Chen, X. & Cheng, Y. (2000) *Urbanization and Its Consequences.*

Smith, D. M. (1977) *Human Geography – A Welfare Approach*, Edward Arnold Publishers Ltd, London.

World Health Organization. (2000) *The World Health Report 2000: Health Systems: Improving Performance*. World Health Organization, Geneva.

11 IDD (Iodine Deficiency Disorders) Scenario in Jammu Province of J&K, India

Rajiv Kumar Gupta and Sunil K Raina

Introduction

The erstwhile princely state of Jammu and Kashmir has currently three administrative divisions viz. Jammu, Kashmir and Leh. Jammu division is comprised of ten districts which include Jammu, Udhampur, Samba, Kathua, Reasi, Rajouri, Poonch, Ramban, Doda and Kishtwar. Except Jammu, Samba and to some extent Kathua districts, most of terrain of Jammu division is mountainous and includes the Pir Panjal Range which separates it from Kashmir valley. Chenab River is the principal river of the Jammu division. Jammu city, also known as the city of temples, is the largest city in Jammu division and is the winter capital of the state. Major religion of Jammu division is Hinduism (62%) and Islam (36%). The Pir Panjal Range, the Trikuta Hills and the low-lying Tawi River basin add diversity to the terrain of Jammu division.

Micronutrient Deficiency

Nutrition security, a fundamental right, ensures optimal utilization of human resources for the overall progress and development of a society and the nation. The deficiencies of micronutrients, also labelled as "hidden hunger" still is a major health problem across the globe but more so in developing nations including India. Five leading micronutrients – iodine, iron, folic acid, vitamin A and zinc – are the cause of concern for global public health planners.

Iodine Deficiency in India

The entire population in India is prone to IDD. An estimated 350 million people in India do not consume adequately iodized salt and, therefore, are at risk for IDD. As per a survey conducted in 325 districts surveyed in India, 263 were found to be IDD-endemic. The survey also pointed out that the household level iodized salt coverage in India was 91% with 71% households consuming adequately iodized salt.

DOI: 10.4324/9781003329459-11

Table 11.1 Earliest studies on Goitre prevalence

Region	Incidence (%)	Number of persons surveyed	Year of survey	Source of information
Kashmir				
Karakoram	90	Not stated	1945	*Indian Med. Gaz.* **80**, 606
Uttar Pradesh				
Dehra Dun	32	554	1945	State public-health department
Bareilly	26	133	1947	*Indian Med. Gaz.* **82**, 23
Bihar				
Purnia District	50	3 villages	1952	State public-health department
East Punjab				
Shiwalak Range	32	5042	1952	State public-health department
Shiwalak Range	37	1337	1952	State public-health department

Source: Authors.

Sub Himalayan Goitre Belt

Over the years, it has been known now that the southern slopes of the Hindu Kush and the Himalayas, covering a distance of over 1,500 miles (2,400 km) and comprising the northern parts of Kashmir, Punjab, Uttar Pradesh (formerly the United Provinces), Bihar, Bengal, and Assam are probably the world's most classical areas of endemic goitre, one of the commonest outcomes of Iodine deficiency. Studies conducted in these areas as far back as the middle of the last century would record that in some villages it was hard to find a man, woman or child not suffering from goitre pointing towards profound iodine deficiency. One of the pioneers on Goitre found that 60% of infants still at breast had goitre in this Himalayan belt. Table 11.1 given below provides some details from one of the earliest study in this regard:

Aetiology of Iodine Deficiency in Sub Himalayan Goitre Belt

Researchers have been working on the aetiology of goitre in this region. The research on iodine deficiency evolved from infection with an organism of the coli-group being a causative factor to drinking of hard water containing excessive amounts of calcium salts. The occurrence of iodine deficiency in Punjab was attributed to excessive intake of fluorine, where a close association between the incidence of endemic goitre and fluorosis was found. However, the key of iodine deficiency is the in Himalayan Goitre is deficiency of Iodine in food items. Although the fact remains that the immediate cause of simple goitre is failure of the thyroid gland to obtain a supply of iodine sufficient to maintain its normal structure and function, but the underlying focus of research in this regard has been on geographical determinants of iodine deficiency.

Therefore the Iodine deficiency disorders (IDD) have largely been linked to deficiency of iodine in the soil. This deficiency has been linked to events like glaciations, flooding, deforestation and changing river course, leading to a constant leaching of iodine present in the top soil. The resultant crops grown on this iodine deficient soil are iodine deficient which means low iodine availability for both livestock as well as humans beings. The iodine content of plants grown in iodine deficient soils may be as low as 10 micrograms/kg compared to 1 mg/kg dry weight in plants growing in iodine replete soil. It accounts for the occurrence of severe iodine deficiency in Asian people living within systems of subsistence agriculture in flooded river valleys of India, Myanmar and Bangladesh.

Adverse Effects of Iodine Deficiency

The recommended daily allowance for iodine is 100–150 micrograms. Evidence generated over the years has demonstrated that iodine deficiency leads to a spectrum of disorders starting from intrauterine life and extending through childhood into adult life with grave health and social problems.

IDD constitute the single largest cause of preventable brain damage worldwide which results in learning disabilities and psychomotor impairment. Lower intelligence quotient (IQ) by 13.5 points has been reported in children living in iodine deficient areas. Two billion people in the world are at risk of iodine deficiency disorders due to suboptimal iodine intake. Globally, India has the largest number of children born vulnerable to iodine deficiency. Iodine deficiency has been known since ancient times.

The National IDD Control Programme

It was first documented by McCarrison in 1908 in India. It was followed by studies of Stott et al. and Ramalingaswami .The pioneer study conducted by Ramalingaswami and his team in Kangra valley, Himachal Pardesh from 1956 to 1972 clearly demonstrated the effectiveness of iodized salt in reducing goiter prevalence in the region. It was in view of the findings of Kangra valley study that Government of India established the National Goitre Control Programme (NGCP) in 1962 with the objective to identify the goitre endemic regions of the country and supplement the intake of iodine to the entire population in these regions . In 1992, the programme was redesignated as National IDD control Programme (NIDDCP) in view of the evidence that "IDD is a spectrum of diseases". The primary aim of this program was to replenish the iodine deficient food consumed by Indians.

The success of NIDDCP was realised through the objectives of NIDDCP, which were inbuilt in it to analyse the outcomes. Large-scale surveys were conducted under the program to assess the magnitude of IDD, supply of iodated salt in place of common salt, laboratory monitoring of iodized salt and urinary iodine excretion and health education and advocacy. After the initiation of the program, a marked reduction has been observed in the spectrum of symptoms produced by iodine deficiency like goitre, stunted physical growth, mental retardation, lassitude, impaired hearing, speech and movement. Improvement was also seen among women exposed to iodine deficiency and thereby reducing the chance of miscarriages, stillbirths and decreased fertility. Better iodine supply to the foetus from the mother could improve the birth parameters of Indian children.

Dynamic evolution of NIDDCP in India is an example of successful translation of research to policy and to the programme. The surveys conducted by NIDDCP in the country have revealed that out of the 414 districts surveyed so far, 337 districts were found to be endemic i.e. where prevalence of IDDs was more than 5%.

Control of IDD in India has been a public health success story. NFHS 4 reported that 93.1% of the households were using iodized salt in comparison to 71% reported by coverage evaluation survey of 2009. The progress attained in the usage of adequately iodized salt at household level since NFHS 2 has been depicted in Figure 11.1.

Iodine Deficiency in J& K

The state of Jammu and Kashmir, for it being geographically located in the sub-Himalayan goitre belt, iodine deficiency disorders have been use for concern since the middle of last century. Studies conducted in this regarded and reproduced here as Table 11.1 is a proof to this. Some of the later studies in this regard have revealed that Iodine deficiency continues to be a health concern in Jammu and Kashmir. For example, for the state of Jammu and Kashmir, the national family health survey (NFHS 4) has reported that 95.5% of the households were using iodized salt. In the urban region, it was 99.5% while it was 93.5% for the rural areas. The National Family Health Survey 2015–16 (NFHS-4) is the fourth in the National Family Health Survey series. NFHS provides information on population, health and nutrition for India and each State / Union territory. NFHS-4 moved further and for the first time, provided district-level estimates for many important indicators.

The results are a marked improvement over NFHS 3 which reported 90.5% of the households using iodized salt .As per NIDDCP, all the 22 districts surveyed in the state of Jammu and Kashmir have been found to be endemic (prevalence > 5%). Among the earlier studies conducted in the state, Zargar et al. reported a goitre prevalence of 52.08% in school children in Kashmir valley. Urinary iodine excretion was 41.85 + 2.52 in randomly selected sample of students.

In the Jammu region, Bhat et al. 2008 reported a goitre prevalence rate of 11.98%. Median urinary iodine excretion was 96.5 microgram/litre (29.0–190 microgram/litre). Forty-nine percent of subjects had biochemical iodine deficiency with 6.7% having moderate and 42.53% mild iodine deficiency. 74.47% of households were consuming powdered salt with 98.17% of the powdered salt samples having an iodine content of >15 ppm. The authors reported district wise goitre prevalence as follows: Jammu (3.5%), Kathua (10.3%), Poonch (13.4%), Doda (21.2%), Rajouri (10.8%) and Udhampur (14%). When this study was conducted, Jammu division had six districts.

In an another study conducted in Jammu region by Gupta et al., goitre prevalence was reported as 21.7%,26.25% and 26.75% in the districts of Jammu, Udhampur and Samba respectively. But the limitation of the study was that neither urinary excretion

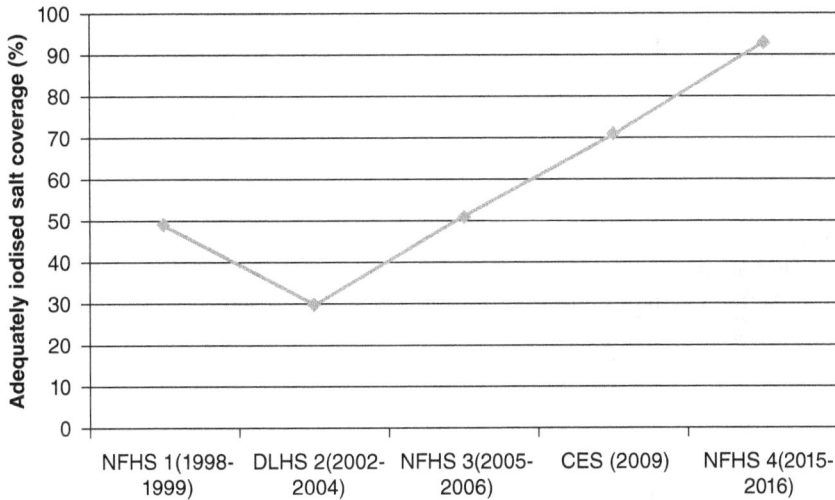

Figure 11.1 Progress in adequately iodized salt at household level in India as confirmed using salt testing kits.

of iodine nor the adequacy of iodine in salt at household level was done. In another study in two borders districts of Rajouri and Poonch, total goitre rate was reported to be 18.87% and 9.70% respectively. The results have shown a paradoxical situation in these two districts as despite as despite TGR being on higher side, all the analysed salt samples collected from the households were found to be adequately iodized. The authors have tried to explain this to the dietary habits of the people who consume foods like soyabean, millets, cabbage, cauliflower, etc., which have naturally occurring goitrogens. Another probable reason could be leaching of iodine by erosion of soil in the mountainous areas of both these districts.

In a yet another study conducted in rural area of Jammu district in children aged 6–12 years, authors reported a goitre prevalence of 12.1%. The prevalence of goitre was higher in children consuming non-iodized salt, those from lower income groups and whose mothers were non-literate. The results also revealed high prevalence of goitre among children who were consuming goitrogenic vegetables (cabbage, cauliflower, etc.) in comparison to the children who were not consuming goitrogenic vegetables.

In view of the literature available on total goitre rate in various districts of Jammu region, the results clearly show that iodine deficiency remains a major public health problem in the region. To tackle the current scenario, future agenda should include all the points discussed below. Health being a state subject as per constitution of India, it is imperative that due focus be given to state level activities. State specific action plans to control IDD with all stake holders playing a key role is likely to further the agenda of IDD in the country. It is pertinent to enhance the need for capacity building, institutionalization of IDD research and advocacy in academic institutes at state level. Wide inequalities still exists in iodized salt coverage across districts in the state, across rural urban divide, across socio-economic strata and amongst marginalized populations. There are also considerable variations in the level of iodization, high moisture content of salt and use of improper packaging material leading to suboptimal iodine content of salt at consumer level. All these issues need to be addressed.

Initiatives to Prevent Iodine Deficiency in J&K

A "mission approach" with greater coordination amongst all stakeholders of IDD control efforts in India is required. The idea is to mainstream IDD control in policy making, devising State specific action plans to control IDD, strict implementation of Food Safety and Standards (FSS) Act, 2006, addressing inequities in iodized salt coverage (rural-urban, socio-economic), providing iodized salt in Public Distribution System, strengthening monitoring and evaluation of IDD programme and ensuring sustainability of IDD control activities are essential to achieve sustainable elimination of IDD in India.

Concerted efforts are required so that every child born is protected from brain damage due to inadequate iodine uptake. In this context, mechanisms need to be evolved to reach vulnerable subpopulations like pregnant women and newborn children. Urinary iodine monitoring of pregnant women should be incorporated as a vital component of NIDDCP in the country. To ensure universal and optimal iodine nutrition to the left over and marginalized populations, we need to target Public Distribution System (PDS) along with ICDS (Integrated Child Development Services) and Mid day Meal programme. Strengthening of state IDD cells is equally important both in terms of human resource as well as technical capacity.

Recently a Management Information System (MIS) was launched by the salt department which deploys state of art web technologies to ensure real time flow of information related to salt production and quality and to increase efficiency in monitoring and controlling flow of iodized salt in India. Ninety-eight percent of the total iodized salt production in the country is from private sector which in itself speaks volumes about responsible participation by private sector in a major national public health programme.

Sustained advocacy at national, state and district level is the need of the hour to ensure political commitment and prioritization of the USI programme. Innovative education, communication and social mobilization strategy is required to generate a strong consumer demand for adequate salt.

References

Bhat IA, Pandit IM, Mudassar S. (2008) Study on prevalence of iodine deficiency disorder and salt consumption in Jammu region. *Ind J Comm Med*;33(1):11–14.

Bleichrodt N, Born MP. (1994) A meta-analysis of research on iodine. And its relationship to cognitive development. In: Stanbury JB, editor. *The damaged brain of iodine deficiency - Cognitive behavioral, neuromotor, educative aspects*. New York: Cognizant Communication Corporation; pp. 195–200.

De Benoist B, McLean E, Andersson M, Rogers L. (2008) Iodine. Deficiency in 2007: global progress since 2003. *Food Nutr Bull*;29:195–202.

Department of Health. Act 4755. (1947) *Part IV Directive Principles of State Policy.* New Delhi: Ministry of Health and Family Welfare, Government of India.

Directorate General of Health Services (DGHS). (2003) *Policy guidelines on national iodine deficiency disorders control programme.* New Delhi: DGHS, Ministry of Health & Family Welfare, Government of India; pp. 1–10.

Gupta RK, Langer B, Raina SK, Kumari R, Jan R, Rani R. (2016) Goiter prevalence in school-going children: A cross-sectional study in two border districts of sub-Himalayan Jammu and Kashmir. *J Family Med Prim Care*;5(4):825–828

Gupta RK, Verma AK, Jamwal DS, Mengi V. (2012) Goiter prevalence in school children in three districts of Jammu region. *J Med Sci*;15:28–31.

ICCIDD, UNICEF, WHO. (2007) *Assessment of iodine deficiency disorders and monitoring their elimination: a guide for programme managers.* Geneva: World Health Organization.

Iodine Network. Global scorecard 2010. Available from: http://www.iodinenetwork.net/documents/scorecard-2010.pdf, accessed on February 21, 2018.

Karmarkar MG, Pandav CS, Yadav K, Kumar R. (2018) *'Mission Approach' to achieve sustainable elimination of iodine deficiency disorders (IDD) in India.* World Journal of Endocrine Surgery. Jaypee Brother Medical Publishers (P) Ltd. Available from: http://www.jaypeejournals.com/eJournals/ShowText.aspx?ID=3226&Type=FREE&TYP=TOP&IN=_eJournals/images/JPLOGO.gif&IID=248&isPDF=YES, accessed on March 1,

McCarrison R. (1909) Observations on endemic cretinism in the Chitral and Gilgit valleys. *Proc R Soc Med*;2:1–36.

Micronutrient Initiative. (2009) *Investing in the future – a united call to action on vitamin and mineral deficiencies.* Global Report. New Delhi: Micronutrient Initiative; Available from: http://www.unitedcalltoaction.org/documents/Investing_in_the_future.pdf, accessed on February 1, 2018.

Pandav CS, Ansari MA, Sundaresan S, Karmarkar MG. (2008) Salt for 57. Freedom and Iodized Salt for Freedom from Preventable Brain Damage. 5th ed. Indian Coalition for control of Iodine deficiency Disorders (ICCIDD); September. Available from: http://www.iqplusin.org/Iec.htm, accessed on March 11, 2018.

Pandav CS, Kochupillai N, Karmarkar MG, Nath LM. (1986) Iodine deficiency disorders in India: review of control measures. *Indian Pediatr*;23:325–329.

Pandav CS, Yadav K, Srivastva R, Pandav R, Karmarkar MG. (2013) Iodine deficiency disorders (IDD) control in India. *Indian J Med Res*;138:418–433.

Ramalingaswami V. (1953) The problem of goitre prevention in India. *Bull World Health Organ*;9:275–281.

Singh G, Kaur G, Mengi V, Raina SK. (2013) Differentials in prevalence of goiter among school children(6-12 years) of age in rural NorthWest India. Pub Health Res;3(3):79–84.

Stott H, Bhatia BB, Lal RS, Rai KC. (1931) The distribution and cause of endemic goitre in the United Provinces. *Indian J Med Res*;18:1059–1085.

UNICEF. Coverage Evaluation Survey 2009. (2018) *All India Report. Ministry of Health and Family Welfare.* New Delhi: Government of India. Available from: http://www.unicef.org/india/health.html, accessed on February 11

Zargar AH, Shah JA, Masoodi SR, Laway BA, Shah NA, Mir MM. (1997) Prevalence of goiter in school children in Baramulla(Kashmir Valley). *Indian J Pediatr*;64(2):225–230.

12 Extension of Vector-Borne Diseases into J&K State of India

Sunil Kumar Raina and Rajiv Kumar Gupta

Background

The state of Jammu and Kashmir in North West of India is home to some of the most beautiful valleys including the world famous Kashmir Valley [Jammu and Kashmir – Wikipedia]. The state (J&K) has been divided into three administrative geographic units; Jammu, Kashmir and Ladakh. The Himalayas divide the Kashmir valley from Ladakh while the Pir Panjal range, which encloses the valley from the west and the south, separates it from the Jammu division and the Great Plains of northern India. [Jammu and Kashmir – Wikipedia.] The state witnesses a range of elevations bringing diversity to its biogeography. While the Jhelum River is the only major Himalayan river flowing through the Kashmir valley, Ind us, Tawi, Ravi and Chenab are the major rivers flowing through the state. [Jammu and Kashmir – Wikipedia.]

Like the diversity in its biogeography, the climate of Jammu and Kashmir varies greatly owing to its rugged topography. [Jammu and Kashmir – Wikipedia] The climate in the south around Jammu is typically monsoonal with Jammu city witnessing temperature of more than 40°C (104°F) in May and June, whilst in July and August, very heavy though erratic rainfall occurs with monthly extremes of up to 650 millimetres (25.5 inches). Srinagar, the main city of Kashmir division receives as much as 635 millimetres (25 inches) of rain, with the wettest months being March to May with around 85 millimetres (3.3 inches) per month. In summer in Ladakh, days are typically warm at 20°C (68°F), but with the low humidity and thin air, nights are still cold.

Vector-Borne Diseases

Malaria and Dengue remain the most commonly known vector-borne diseases in the region. The list of vector-borne diseases also includes Lymphatic Filariasis, Kala-Azar, Japanese Encephalitis and Chikungunya. Out of this list, malaria and dengue are the most rapidly spreading mosquito-borne diseases and of significant importance in terms of both mortality and morbidity. [National Vector Borne Disease Control Programme (NVBDCP)] Both these diseases raise large-scale public health concerns in view of rapidly occurring industrialization and globalisation around us.

DOI: 10.4324/9781003329459-12

India's public health has seen significant improvement in mother and child health over the past three to four decades[Raina SK et al. 2018]. This improvement has seen some decline in its pace in the past few years. In addition to this decline in pace in reforms, there is appreciable threat of rise in mortality attributable to diseases such as dengue fever, malaria, tuberculosis and pneumonia which continue to plague India. One of the most important reasons for this threat is an increase in drug resistance across the length and breadth of this country.

Malaria is a protozoan parasitic infection caused by Plasmodium species. The major Plasmodium species infecting humans are *Plasmodium falciparum* (*P. falciparum*), *Plasmodium vivax* (*P. vivax*), *Plasmodium ovale* and *Plasmodium malariae* [Dhiman et al. 2010]. Similar to Malaria, dengue a mosquito-borne (*Aedes aegypti*) disease is caused by single stranded RNA viruses of four distinct serotypes (DEN-1, 2, 3 and 4). Again like malaria, the typical transmission cycle in dengue also follows the human-vector-human cycle. A common cause for acute undifferentiated fevers (AUF) is the concurrent infections of malaria and dengue which can occur simultaneously in an individual. AUF poses a diagnostic and therapeutic challenge to health care workers, particularly in resource limited settings. Further because of the similarities in the clinical characteristics between malaria and dengue, diagnosis of malaria and dengue co-infections might be either misdiagnosed or misinterpreted as mono-infections [Selvaretnam et al. 2016].

Climate Change

As per The United Nations Framework Convention on Climate Change (UNFCCC), climate change is a change of climate which is attributable directly or indirectly to human activity, alters the composition of the global atmosphere and is in addition to natural climate variability observed over comparable time periods. [Intergovernmental Panel on Climate Change (IPCC) (2007) Summary for policymakers. In: Climate Change 2007.]

A pointer to climate change is the rise in temperature, which has become more apparent since 1990. As per the Fourth Assessment Report of IPCC 2007 based on six scenarios, the rise in temperature is projected from best estimated figure from 1.8°C to 4°C by the year 2099 relative to 1980–99 (IPCC 2007). [Intergovernmental Panel on Climate Change (IPCC) (2007) Summary for policymakers. In: Climate Change 2007.] The report states that the sea level is expected to rise up to 0.59 m by the year 2100. It is being estimated that tropical cyclones (typhoons and hurricanes in future will tend to become more intense, with larger peak wind speeds and more heavy precipitation. The report concludes that climate change is likely to expand the geographical distribution of several vector-borne diseases, including malaria, dengue and leishmaniasis to higher altitudes (high confidence) and higher latitudes and extend the transmission seasons in some locations.

The expected impact of climate change across India is not uniform. While an overall decrease in the number of rainy days over major part of the country is expected, this decrease is greater in the western and central parts (by more than 15 days) while near Himalayan foothills, the number of rainy days may increase by 5–10 days [Dhiman RC et al. 2010].

The Logic of Impact of Global Warming on VBDs

A study of the epidemiology of VBDs reveals that all the VBDs are climate sensitive. The reason being, that, the pathogens involved in VBDs have to complete part of their development in particular species of an insect vector. The insects being poikilothermic, development of parasite in their body are affected by climatic conditions. For example, at a temperature of 16°C, it will take 55 days for completion of sporogony of *P. vivax* while at 28°C, the process can be completed in seven days. As the temperature rises, the chance of survival of mosquitoes reduces and at 40°C, their daily survival becomes zero. [Craig MH et al. 1999; Martens WJM et al. 1999] Rainfall is another important climatic factor for mosquito breeding. An increase in rainfall can increase the breeding sites of mosquitoes. Heavy rainfalls may impact adversely also by wiping out the breeding grounds. Similarly dry conditions can either eliminate or create several new breeding habitats in large water bodies such as lakes and rivers. Rainfall also increases relative humidity (RH) and thereby modifies temperature affecting the longevity of mosquitoes and therefore transmission of the disease [Molineaux L et al. 1980].

Therefore, climatic conditions largely decide the distribution, degree of endemicity and epidemicity of a vector-borne diseases in a particular geographical area. Geographical locations with conditions favourable to vector growth may experience transmission of disease throughout the year, while areas generally experiencing colder months, transmission are seasonal. In addition, developmental activities like construction of dams and increase in irrigation channels, human migration, agriculture practices, education, human behaviour and economic conditions also influence transmission of VBDs.

Research work by investigators across the world has suggested some models reflecting on the epidemic potential of malaria. Jetten et al. in 1996 have projected two to five times changes in epidemic potential for *P. falciparum* malaria with 2–4°C increase in temperature; highest changes are projected for high altitudes [Jetten TH et al. 1996]. These projections are based on distribution and vectorial capacity of malaria vectors.

Situation Analysis of VBD's in India

Vector borne diseases are endemic to almost all parts of India with Indo-gangetic plains generally presenting with higher incident cases. However data on VBD's in recent years is pointing to changing trends. This trend is apparent in the monthly incidence reports of malaria from different states. The reasons for these changing trends could be because of changes in climate across geographies in India favouring vector borne diseases. It is being established that the vulnerability of northern states of Jammu and Kashmir, Himachal Pradesh, Punjab, Haryana, Uttarakhand and the north-eastern states of India to climate change may favour occurrence of VBD's in these regions. The climate of the southern states of India, such as Karnataka, Kerala, Tamil Nadu and Andhra Pradesh as well the states of Orissa stays almost the same throughout the year, thereby making them less vulnerable to climate change as the climatic conditions are already suitable for malaria transmission almost throughout the year [Dhiman RC et al. 2010].

Studies have been undertaken by National Institute of Malaria Research by conducting baseline surveys across all states of India to build a map showing transmission windows (TW) of malaria across these states. The map generated TW based on different months of the year across different states of India. The idea was based month wise recording of temperature and thereby calculating the TW on the basis of months providing the minimum required temperature of 19°C and relative humidity of 55% for mosquito growth. The map shows that in Rajasthan, Punjab, Haryana, Himachal Pradesh, Jammu & Kashmir and Meghalaya, the TW is open for only 4–6 months. In states like Gujarat, Orissa, West Bengal and southern states the TW for malaria is open for 10–12 months. Therefore, the projected rise in temperature is more likely to affect the TWs in Punjab, Haryana, Jammu and Kashmir, Himachal Pradesh and the north-eastern states. On the other hand, the states of Orissa, AP and Tamil Nadu are expected to register a reduction in TWs due to high temperatures [Dhiman RC et al. 2010]. Importantly, however, in such situations mosquitoes may adapt and rest in micro-environments favourable for their survival. A study conducted in Rajasthan, according to which the adult mosquitoes prefer to rest in 'tankas' to avoid high temperatures and increase their survival in adverse climatic conditions.

In a similar study, Bhattacharya et al. have projected that the temperature and RH for year 2050 in Orissa, West Bengal and southern parts of Assam will still remain malarious, while the transmission windows will open up in Himachal Pradesh and north-eastern states [Bhattacharya S et al. 2006]. Essentially therefore, the duration of transmission windows is likely to widen in northern and western states and shorten in the southern states.

But, importantly, malaria transmission is not entirely dependent on temperature and relative humidity only. The dynamics of malaria transmission are also decided by factors like agricultural practises, deforestation, socio-economic conditions, pre-existing health status, quality/availability of public health care and types of intervention measures undertaken. Therefore in contrast to other vector-borne diseases like Dengue for example, the projection of change in malaria transmission based on temperature and RH alone may not hold true.

Dengue in comparison to malaria is currently more commonly distributed across the southern part of India, although dengue epidemics have been reported in North as well. The main limiting factors responsible for dengue transmission are the temperature, rainfall and RH [Ansari MA et al. 1998; Sharma SK et al. 1998]. One other factor that is being realised as important in dengue epidemics is the water management associated with urbanisation coupled with lack of surveillance and vector control [Gubler DJ 1997, 1998; Gupta et al. 2005, 2006]. Rapid urbanisation has fuelled a scarcity of water. This leads to water storage and stored water serves as an excellent breeding ground for aedes aegypti – the mosquito vector for dengue.

Now, with rapid urbanisation picking in most part of India including Jammu & Kashmir and the projected rise in temperature, dengue is also likely to extend towards northern parts of India. It is being projected that under climate change scenario, overall, with 4°C rises in temperature, dengue transmission may be two to five times more with new transmission areas in northern sub-Himalayan region like Jammu & Kashmir and Himachal Pradesh.

In contrast to malaria and dengue, influence of climatic changes on Visceral leishmaniasis may be slow to occur. Leishmaniasis is caused by Leishmania parasite and phlebotomine sand flies acts as the transmission vector. The disease like other VBDs is climate sensitive as the preferred breeding ground for the sand fly vectors is the alluvial soil with high sub-soil water table and temperature ranging from 7°C to 37°C with a RH above 70% in India [Napier LE 1926].

The state of Himachal Pradesh, in Sub-Himalayan region of India was free from kala-azar till 1984. However since then pockets of infection have been noticed [Datta U et al. 1984] After a gap of about 20 years, cases started emerging since 2004 and now cases have been reported from Shimla, Kinnaur and Kullu districts [Mahajan SK et al. 2004, Sharma NL et al. 2009]. Similarly Garhwal region of Utt-arakhand state started reporting kala-azar cases (unconfirmed origin) since 1984. These reports point to possibility of occurrence of these diseases in other Sub-Himalayan states like the Jammu & Kashmir.

Situation Analysis of J&K

Out of the six major vector-borne diseases in India, only malaria, dengue and to some extent Chickungunya are known to have recorded their presence in Jammu and Kash-mir. Malaria in J&K state is confined to Jammu province and a portion of district Bar-amulla and Kupwara in Kashmir province. The three districts of Jammu and Kashmir viz, Jammu, Kathua and Udhampur have been recording malaria continuously. The Kashmir province of J&K has continuously been recording an Annual Parasite Inci-dence of <0.1 [Gupta RK et al. 2016]. The reason could lie in the fact that the Jammu province of J&K has more physiographical resemblance to the rest of North India than the Kashmir province of the state. However, dengue was hypoendemic in the Jammu province despite the fact that the climate of Jammu is favourable for growth and spread of virus. Study on district wise distribution of seropositive samples of Dengue showed a clear trend towards more number of cases occurring from the three districts mentioned above in comparison to the rest of districts of Jammu province which have a hilly terrain. The study also observed a rise of Dengue cases in the year 2013, which was partially attributed to the rapid unplanned urbanization with unchecked construction activities and poor sanitation facilities contributing fertile breeding grounds for mosquitoes [Sudhan SS et al. 2016]. Another study conducted in Jammn region about serosurveillance in 2016 reported 17.82% and 18.2% positiv-ity for dengue and chickungunya respectively although no case of Zika virus infec-tion was reported by the authors [Sudhan SS et al. 2017].

However a report published in a newspaper in 2015 raises fresh concerns. It said that four fresh cases of dengue were detected taking the total number to 17 in the entire Jammu region. The concern was raised because a youth in Basohli area of Kathua district had tested positive for dengue. Basholi located at 32.50°N 75.82°E, has an average elevation of 460 metres (1,509 feet) [Basholi – Wikipedia]. Situated on the right bank of Ravi River, Basholi has become popular for the Thein dam which has made it almost landlocked.

Travel of the population from and to dengue endemic neighbouring states may be responsible for some of these imported cases. Generally speaking the Himalayan region above the height of 500 meters has been found completely free of the main dengue vector, *Aedes aegypti*. Importantly however, presence of *Aedes albopictus* has been reported in other Himalayan areas [Raina S et al. 2018]. Therefore there is an apparent geographical shift due to the modifying effects to the ecosystem associated with change in temperature (particularly night time minimum temperature), precipitation and peak relative humidity.

Conclusions

The change in climate across geographies is expected to extend the presence of vector borne diseases in India. The Jammu and Kashmir state of India with currently less common presence of Vector borne diseases may start recording a higher presence in future with this climate change.

References

Ansari MA, Razdan RK. (1998) Seasonal prevalence of Aedes aegypti in five localities of Delhi, India. *Dengue Bull*; 22:28–32.
Basholi - Wikipedia. Available online at https://en.wikipedia.org/wiki/Basholi. Last accessed: 20-05-18
Bhattacharya S, Sharma C, Dhiman RC, Mitra AP. (2006) Climate change and malaria in India. *Curr Sci*; 90:369–375.
Craig MH, Snow RW, DA Sueur LE. (1999) Climate based distribution model of malaria transmission in sub-Saharan Africa. *Parasitol Today*; 15:105–111.
Datta U, Rajwanshi A, Rayat CS, Sakhuja V, Sehgal S.(1984) Kala-azar in Himachal Pradesh: a new pocket. *J Assoc Phys India*; 32:1072–1073.
Dhiman RC, Pahwa S, Dhillon GPS, Dash GP. (2010) Climate change and threat of vector-borne diseases in India: are we prepared? *Parasitol Res*; 106:763–773.
Gubler DJ. (1997) Dengue hemorrhagic fever: its history and resurgence as a global health problem. In: Gubler DJ, Kuno G (eds) *Dengue hemorrhagic fever*. CAB International, New York, pp. 1–22.
Gubler DJ. (1998) Dengue and dengue hemorrhagic fever. *Clin Microbiol Rev*; 11:480–496.
Gupta E, Dar L, Narang P, Srivastava VK, Broor (2005). Serodiagnosis of dengue during an outbreak at a tertiary care hospital in Delhi. *Indian J Med Res*; 121:36–38.
Gupta E, Dar L, Kapoor G, Broor S. (2006) The changing epidemiology of dengue in Delhi. *Virol J*; 3:92–98.
Gupta RK, Raina SK, Shora TN, Jan R, Sharma R, Hussain S. (2016) A household survey to assess community knowledge and practices on malaria in a rural population of Northern India. *J family Med Prim Care*; 5:101–107.

Intergovernmental Panel on Climate Change (IPCC). (2007b) Summary for policymakers. In: *Climate Change 2007: Impacts, Adaptation and Vulnerability*. Contribution of Working Group II to the Fourth Assessment Report of the Intergovernmental Panel on Climate. Cambridge University Press, pp. 1–93.

Jammu and Kashmir - Wikipedia. Available online at: https://en.wikipedia.org/wiki/Jammu_ and_ Kashmir. Last accessed: 20-05-18.

Jetten TH, Martens WJM, Takken W. (1996) Model simulations to estimate malaria risk under climate change. *J Med Entomol*; 33:361–371.

Mahajan SK, Machhan P, Kanga A, Thakur S, Sharma A, Prasher BS, Pal LS. (2004) Kala-azar at high altitude. *J Commun Dis*; 36:117–120.

Martens WJM, Kovats RS, Nijhof S, deVries P, Livermore MJT, McMichael AJ, Bradley D, Cox J. (1999) Climate change and future populations at risk of malaria. *Global Environ Change*; 9:S89–S107.

Molineaux L, Gramiccia G. (1980) *The Garki project: research on the epidemiology and control of malaria in the Sudan Savanna of West Africa*. WHO, Geneva.

Napier LE. (1926) An epidemiological consideration of the transmission of Kala—azar in India. In Reports of the Kala–azar commission, India, Report No 1 (1924–25). *Indian Medical Research Memoirs*; 4:219–265.

National Vector Borne Disease Control Programme. (NVBDCP). Available online at: www.nvbdcp.gov.in. Last accessed: 20-05-18.

Raina S, Raina RK, Agarwala A, Raina SK, Sharma S. (2018) Co- infections as an etiology of acute undifferentiated febrile illness among adult patients in the Sub Himalayan region of North India. *J Vector Borne Dis;* 55:130–136.

Raina SK, Galwankar SC, Bhat R, Bodhankar U, Prabhoo R, Mishra SK. (2018) Organised medicine: need for a guild of association. *J Global Infect Dis*; 10:35–36.

Selvaretnam AP, Sahu PS, Sahu M, Ambu S. (2016) A review of concurrent infections of malaria and dengue in Asia. *Asian Pac J Trop Biomed*; 6(7):633–638.

Sharma SK. (1998) Entomological investigations of DF/DHF outbreak in rural areas of Hissar District, Haryana. *India Dengue Bull*; 22:36–41.

Sharma NL, Mahajan VK, Ranjan N, Verma GK, Negi AK, Mehta KS. (2009) The sandflies of the Satluj river valley, Himachal Pradesh (India): some possible vectors of the parasite causing human cutaneous and visceral leishmaniasis in this endemic focus. *J Vector Borne Dis*; 46:136–140.

Sudhan SS, Sharma M, Gupta RK, Sambyal SS. (2016) Sero-epidemiological trends of dengue fever in Jammu province of J&K State. *Int J Med Res Health Sci;* 5(8):1–6.

Sudhan SS, Sharma M, Sharma P, Gupta RK, Sambyal SS, Sharma S. (2017) Serosurveillance of Dengue, Ckickungunya and Zika in Jammu, A sub-Himalayan region of India. *J Clin Diag Res;* 11(11):5–8.

13 Extreme Weather Events

Floods and Disease Risks in Kashmir (Jammu & Kashmir). A Case Study of Srinagar Floods

Bupinder Zutshi

Introduction

Several studies by the scientific community have indicated that the Earth's climate is rapidly changing. The Intergovernmental Panel on Climate Change (IPCC) states that "warming of the climate system is unequivocal," and its conclusion is supported by observations of increases in global air and ocean temperatures, widespread melting of snow and ice, and rising global mean sea level (IPCC, 2007c, p. 5).

Natural phenomena supported by indiscriminate human activities, especially the burning of fossil fuels and changes in land use patterns, are considered to be the main reasons for the climatic changes observed since the mid-20th century (IPCC, 2007c).

Extreme weather events are a global phenomenon. The extreme weather events have impacted both natural and human systems. These systems include water resources, ecosystems, food and forest products, coastal systems and low-lying areas, industry, settlements and societies, and human health, involving significant social, economic and environmental consequences (Stern, 2006; IPCC, 2007c).

Periodic assessments of vulnerabilities to extreme climate changes, especially precipitation variations over several regions are necessary to inform communities facing such extreme events, so that they are ready to support the processes of adaptation. Information and knowledge about the health impacts of extreme weather change is growing rapidly and is increasingly being shared around the world. Several recent publications examining the effects of climate change at the global, national and regional levels have highlighted the health impacts of climate change on populations (Mcmichael et al., 2003; Berner et al., 2005; Menne and Ebi, 2006; Stern, 2006; Confalonieri et al., 2007).[1]

It is widely recognized that efforts must focus on assessing current and future health vulnerabilities in order to identify actions to help those affected, especially the most vulnerable. Many harmful health effects of climate change are well established, ranging from the worsening of seasonal allergies and the spread of infectious diseases to the deadly health effects due to the impact of heat waves, floods, and drought, among others. Existing evidence also suggests that climate change already is tied to heart problems, from the dangerous effects of air pollution, including from climate fuelled to that of stress, a known risk factor for heart disease.

DOI: 10.4324/9781003329459-13

Recurrent floods across globe is one of the major outcomes of extreme weather events. Flooding of areas is the major source of waterborne diseases that are linked to significant disease burden worldwide. Waterborne diarrhoeal diseases, for example, are responsible for 2 million deaths each year, with the majority occurring in children under 5 years. Climate change-induced flooding and droughts can impact household water and sanitation infrastructure and related health risks. For instance, flooding can disperse faecal contaminants, increasing risks of outbreaks of waterborne diseases such as cholera. In addition, water shortages due to drought can increase risks of diarrhoeal disease.

Water related diseases are classified into four types relating to the path of transmission:

- Waterborne diseases, such as cholera, amoebic dysentery and typhoid, are the diseases that are transmitted through drinking water.
- Water-washed (water-scarce) diseases, such as polio, are diseases where the interruption of the transmission is achieved through proper attention to effective sanitation, washing and personal hygiene.
- Water-based diseases are diseases transmitted by contact with water, e.g. recreational swimming.
- Water vector diseases, such as malaria, are diseases that are transmitted by a vector, such as the mosquito, which needs water or moisture in order to breed. Prevention of transmission is through a vector control.

Waterborne diseases can have a significant impact on the economy. People infected by a waterborne disease are usually confronted with related costs and financial burden. This is especially the case in less developed countries. The financial losses are mostly caused by e.g. costs for medical treatment and medication, costs for transport, special food, and by the loss of manpower.

Occurrence of floods in Kashmir

Kashmir Valley, in Jammu and Kashmir state is highly prone to floods, due to its physiographic and geologic structure. Kashmir Valley is a deep, asymmetric basin, delineated from the adjoining mountain systems, on the basis of drainage network and its catchment basin. The valley of Kashmir includes, all land lying within the water divides, formed by the Pir-panjal Mountain ranges, the north Kashmir ranges and the great Himalayan ranges. These mountain ranges encircle the great synclinal trough, occupied by the Jhelum River (Raza Moonis, 1978). Average height of Kashmir Valley is about 1850 meters above sea level, but the surrounding mountains, which are generally snow clad, rise up to 3,000–4,000 meters, above sea level (Census of India, 1988). Like all sedimentary basins the valley has a combination of depositional and erosional features. The low lying waterlogged areas and Jhelum river channel are subject to, receiving layer after layer of fine silt and coarse gravel that increases, water level of Jhelum river during Monsoon period incessant rainfall. This enhances the flood conditions in the

Table 13.1 Flood events, fatalities, injuries and causalities - 1978–2006

State	Flood events		Fatalities		Injuries, missing, causalities	
	No	*Rank*	*No*	*Rank*	*No*	*Rank*
Jammu & Kashmir	76	13	1365	13	1563	14

Table 13.2 Flood death rates per million population at decennial intervals

State	*1978–87*	*1988–97*	*1998–2006*	*1978–2006*
Jammu & Kashmir	35	146	3	135
India	23	18	15	44

Source: Omvir Singh, Manish Kumar, "Flood Events, fatalities and damages in India from 1978-2006", published online 16th July 2013, Springer Science + Business Media, Dordrecht, 2013.

low-lying areas, as well as in Srinagar city, if bunds/river channel embankment walls breach owing to higher discharge from upper reaches.

The valley of Kashmir has a long history of floods due to excessive water discharge from the tributaries of Jhelum river during strong western disturbances (November to April) and monsoon period (July- September). According to Sir Walter Lawrence "Many disastrous floods are noticed in vernacular history of Kashmir valley" (Walter Lawrence, 1895).

An analysis of human fatalities due to floods in Kashmir valley as compared to the national average depicts more deaths, thereby suggesting lack of disaster management policy and disaster risk-reduction strategies in the state contributing to water-borne disease burden in Kashmir Valley (Tables 13.1 and 13.2). Both tables indicate that Jammu and Kashmir has witnessed 76 flood events during 1978–2006 periods and it ranked 13 among other states of India in terms of number of flood events. The state recorded 1,365 fatalities and 1,563 persons with injuries, missing and other causalities between 1978 and 2006 periods. Flood death rates recorded were significantly high compared to national average for the two decennial periods as well as for the overall selected period indicating least disaster risk reduction strategies adopted by the state of Jammu and Kashmir. Water-borne diseases have shown significant increase during the floods (Discussion with several medical doctors conducted during the field survey).

Occurrence of September 2014 flood in Kashmir Valley

Jammu & Kashmir experienced the worst floods in the past 60 years during first week of September 2014 due to unprecedented and intense rains which was considered as one of the extreme weather events. "The synchronization of movement of westerly winds in the extreme north, with the passage of monsoon disturbances in the lower latitudes caused heavy to very heavy rainfall along the foothills of the Himalaya, and adjoining areas of Jammu & Kashmir" (Kamaljit Ray et al., 2015). Continuous rainfall from 1–6 September2014, with a record of 30 hour of long rainfall from 3rd September 2014, broke the record of many decades. Majority of rainfall stations in Kashmir valley recorded deviation of more than +6,000% rainfall for the four days (3–6 September 2014) as compared the norm of rainfall for the same days recorded from 1970 to 2000 (Table 13.3). Inadequate capacity of the Jhelum river and its tributary rivers to contain within their banks, the high flows brought down from the upper catchment areas following heavy rainfall, lead to flooding of villages. The existing embankment/bunds on river Jhelum in Srinagar city could not contained Jhelum River and as a result of this, nearly 90 breaches were found at several places in and around Srinagar city. These breaches flooded nearly 3/4th of city Srinagar city and its lower level inhabitations.

A total of 1.16 million populations out of 1.27 million population of Srinagar Urban Agglomeration constituting 91 percent population were affected by the September 2014 floods. (NRSC-ISRO, 2014, p54). Out of 227.41 square kilometres areas of Srinagar city agglomeration, 118.75 square kilometres of areas constituting 52% of the Srinagar city areas was completely inundated/submerged. The distribution of fully inundated areas in Srinagar city were, residential (28.11 Sq Kms), agriculture (34 Sq Kms) commercial (32.21 Sq Kms) Open areas (12.39 Sq Kms) and the rest areas including restricted areas (Military areas), hospitals, police stations, educational and areas with administrative buildings.

Table 13.3 Rainfall reported (mm) over stations located in Kashmir Valley during 3–6 September 2014

District	Station	Actual rainfall reported during 3–6 September 2014	Normal rainfall during 3–6 September 2014 **	Deviation from Norm (%)
Anantnag	Kukernag	415	5.5	7449
	Pahalgam (AWS)	217	6.9	3045
Baramulla	Baramula (AWS)	185	1.9	9637
	Gulmarg	366.8	6.5	5543
Kulgam	Kulgam (AWS)	346	9.8	3431
	Qazigund	599.8	5.4	11007
Awantipur	Awantipur (IAF)	208.1	-	-
	Malangpura (AWS)	266	-	-
Shopian	Shopian	335	5	6600
Srinagar	Rambagh (AWS)	140	2.6	5285
Kupwara	Kupwara	131.8	3.1	4152

Source: Kamaljit Ray*, S. C. Bhan and B. K. Bandopadhyay, "The catastrophe over Jammu and Kashmir in September 2014: a Meteorological observational analysis, *Current Science*, Vol. 109, no. 580 3, 10 August 2015.
**Normal Rainfall of these stations have been calculated from 1970- 2000 for the same days.

The flood was termed as one of the worst floods. Dr. Omar Abdullah, the then Chief Minister of Jammu & Kashmir state had to concede that "I had no government for the first 36 hours as the seat of establishment was wiped out."

Disease Risks due to Srinagar Flood 2014: Responses from Field Survey

Floods can potentially increase the transmission of many communicable diseases, and pose other health risks as well. Flooding is associated with an increased risk of infection of the water-borne diseases like; Amebiasis, Gastrointestinal infections, Typhoid, Cholera and vector borne diseases due to mosquito breeding diseases, skin rashes, etc.

A sample survey of flood victim respondents was conducted to assess the disease risks suffered by flood victims in Srinagar city. Four localities from Srinagar city, namely Mehjoor Nagar, Jawahar Nagar, Gogji bagh and Raj bagh were selected for the sample survey. These four localities are residential colonies, who were completely submerged with flood water for a minimum of ten days. In addition to these four localities, Lal Chowk, Central Business District (CBD) areas with large number of shopkeepers and traders was also selected for the survey. Lal Chowk was also completely submerged with flood waters for a minimum of 7–10 days. The survey was conducted during March–May 2016.

Table 13.4, depicts total family size of the selected 40 households for the survey. A total of 94 male and 95 females were covered by the survey. These respondents represented diverse age, occupation and education level groups.

Among the flood households selected for the survey, 15 households were severely affected, 20 households were moderately affected and five households were mildly affected by the 2014 floods. Among the selected respondents five households had complete collapse of house, 20 respondents reported severely/partially damaged houses while ten respondents were shopkeepers/traders whose shops were completely submerged by floods in Lal Chowk CBD area (Table 13.5)

Medical doctors were consulted by the research team to conduct the survey on water borne disease risks suffered by the flood victims. The term waterborne disease is reserved largely for infections that predominantly are transmitted through contact with or consumption of infected water. Medical doctors reported that they

Table 13.4 Sample survey respondent characteristics

Areas/localities	Sample households	Total population in sample households		
		All	Male	Female
Mehjoor Nagar	8	37	19	18
Lal Chowk	10	46	25	21
Jawahar Nagar, Gogji Bagh, Rajbagh	22	96	50	46
All areas	**40**	**179**	**94**	**85**

Source: Field Survey by Research Team, March – May 2016.

Table 13.5 Nature of flood affected in September 2014 (among surveyed households)

Areas/localities	Flood affected families			Nature of damage to property		
	Severe	Moderate	Mild	House collapsed	House submerged	Trade/shops submerged/ collapsed
Mehjoor Nagar	3	5		3	5	
Lal Chowk	5	5				10
Jawahar Nagar, Gogjibagh, Rajbagh	7	10	5	7	15	
All areas	15	20	05	10	20	10

Source: Field Survey by Research Team March – May 2016.

have already seen many cases of diarrhoea, skin allergies and fungus, and were worried that the stagnant waters had creating conditions for the outbreak of serious diseases. Doctors from medical college stated that The chances of cholera, jaundice and leptospirosis spreading was high, as treated drinking was lacking during the early phase of submergence. The research team after consulting several medical doctors advised us to identify the following symptoms to identify the water borne diseases that might have occurred during and after the floods in Srinagar city in 2014. The following Table 13.6 was prepared in consultation with medical doctors to conduct the survey for identifying the disease risks suffered by the selected sample households. Sample households' members were asked if they suffered with any of the identified symptoms during the floods and after the floods for a period on one month. Based on the symptoms identified in Table 13.6, risk of identified water borne disease burden during the Srinagar floods 2014 were identified. Sample households were asked if any of their family member had suffered from any of the following symptoms.

The data collected from the sample survey has been tabulated in Table 13.7, to identify the major waterborne diseases that infected the sample population. Two time frames were selected (August 2013 to August 2014 and September 2014) to measure the burden of waterborne diseases among the selected population in order to compare the water borne disease burden affected due to floods. Only 15% selected sample population had been infected by the identified water-borne diseases during August 2013 to August 2014 period. On the other hand, during the One month flood period (September 2014) waterborne diseases had infected 39% sample population. Giardisis disease was prevalent among 15% sample population, Viral Gastroenterritis was prevalent among 10% sample population and Amebiasis and Cryptosporidiosis had infected by 6% population each. Three percent sample population were also infected by typhoid. Skin rash was major disease caused by floods as 39% of the sample population suffered by skin rash during the floods as compared to 4% sample population.

Biggest challenge faced by the Jammu and Kashmir Government after the September 2014 floods was to meet the demand of people to provide effective health

Table 13.6 Symptoms for identifying water borne diseases suffered by flood victims

Symptoms	Identification of water-borne disease	Cause
Diarrhea, stomach pain, and stomach cramping, bloating, fever	**Amebiasis**	Sewage non-treated drinking water, Fecal matter of an infected person (usually ingested from a flood or an infected water supply)
Flu-like symptoms, Stomach cramps, dehydration, nausea, vomiting, fever, weight loss, increased gas formation.	**Cryptosporidiosis**	Fecal matter of an infected person due to contaminated water, animal manure mixing with water.
Watery diarrhea, vomiting, and leg cramps	**Cholera**	Contaminated drinking water due to water stagnation
Diarrhea, excess gas, stomach or abdominal cramps, and upset stomach or nausea	**Giardisis**	Swallowing recreational water contaminated with Giardia
Diarrhea, vomiting, nausea, cramps, headache, muscle aches, tiredness, slight fever	**Viral Gastroenterritis**	Water, ready-to-eat foods
Characterized by sustained fever up to 40 °C (104 °F), profuse sweating; diarrhea may occur	**Typhoid**	Ingestion of water contaminated with feces of an infected person

Source: 1. Consultation with medical doctors before the conduction of field survey to identify water-borne diseases suffered by flood victims.

Table 13.7 Water-borne disease experienced by the sample families (total respondents covered 179 members)

Name of disease as identified based on Symptoms	Percent respondents experienced the disease before floods (one year period)		Percent respondents experienced the diseases during and after flood 2014	
	Number	Percent	Number	Percent
Amebiasis	5	3	11	6
Cryptosporidiosis	4	2	10	6
Cholera	0	0	0	0
Giardisis	10	6	27	15
Viral Gastroenterritis	5	3	18	10
Typhoid	2	1	4	3
Skin Rashes	3	8	25	14
All (Minus Skin Rash)	8	4	70	39

Source: Field Survey data analysis.

care system as number of patients infected by water borne diseases had increased substantially, while the existing healthcare services were inadequate. Discussion with medical doctors held during the survey substantiated the survey results as they stated high cases of Giardisis, viral Gastroenterritis and Amebiasis. Thus, the floods due to extreme weather conditions have significant impact on the diseases risks.

Stress of Healthcare System

According to the victims of floods the existing healthcare support system like hospitals and private doctor services could not function for several days as the flood had led to closure of some of the premier hospitals in Srinagar, though temporarily. Hospitals also suffered massive damage to machinery as well. Kashmir's private health sector which was mainly housed in Karan Nagar area had suffered massive losses due to complete inundation. The only child care hospital at Sonwar, GB Pant, has been closed due to inundation of the entire area. Sonwar, a low-lying area, was destroyed by floods, damaging its road and power network. Apprehension were that "flood might have damaged hospital's oxygen plant completely," said a senior doctor at the institute. He said hospital authorities were waiting for the flood water to recede in the area for several days, so that the paediatric services could be restored.

The state Health Minister stated "We have a war like situation in the health sector. We are trying to run the health services wherever it is possible". The situation was equally bad at the SMHS hospital located in the downtown. Its ground floor was completely submerged in the flood causing massive damage to some of the hi-tech equipment housed in the hospital. "The hospital which caters to rush from across Kashmir was however re-started partially only after seven days. Another government hospital at Bemina bypass, SKIMS Medical College, which receives patients from entire north Kashmir comprising of three districts, was the first institute that took the hit in the flood, damaging its ground floor. The private health sector also bore the brunt of the devastating floods."

Thus, extreme weather events not only increases the burden of diseases but also impacts the inadequate healthcare services in the under developing and developing countries, which directly has serious consequences on their economies.

Note

1 A list of national impact assessments of climate change can be found in the Health Chapter of the IPCC Fourth Assessment Report (CONFALONIERI et al., 2007). The World Health Organization has also completed, or participated in, a number of assessments of climate change health risks (MCMICHAEL et al., 2003; MENNE and EBI, 2006).

References

Berner, J., Furgal, C., Bjerregaard, P., Bradley, M., Curtis, T., et al. (2005). "Human health". In *Arctic Climate Impact Assessment (ACIA)* (pp. 863–906). Cambridge: Cambridge University Press.
Census of India. (1988). *Regional Divisions of India – A Cartographic Analysis*, Occasional Papers, Series - 1, Volume -Viii, Jammu & Kashmir.

Confalonieri, U., Menne, B., Akhtar, R., Ebi, K.L., Hauengue, M., et al. (2007). "Human health". In M.L. Parry, O.F. Canziani, J.P. Palutikof, P.J. van der Linden, and C.E. Hanson (Eds.), *Climate Change 2007: Climate Change Impacts, Adaptation and Vulnerability. Working Group II Contribution to the Intergovernmental Panel on Climate Change Fourth Assessment Report* (pp. 391–431). Cambridge: Cambridge University Press.

Intergovernmental Panel on Climate Change (IPCC). (2007c). "Summary for policymakers". In S. Solomon, D. Qin, M. Manning, Z. Chen, M. Marquis, et al. (Eds.), *Climate Change 2007: The Physical Science Basis. Working Group, I Contribution to the Intergovernmental Panel on Climate Change Fourth Assessment Report* (pp. 1–18). Cambridge: Cambridge University Press.

Kamaljit Ray, S.C. Bhan and B.K. Bandopadhyay. (2015). The catastrophe over Jammu and Kashmir in September 2014: a Meteorological observational analysis. *Current Science*, 109, 5803.

McMichael, A.J., Campbell-Lendrum, D.H., Corvalan, C.F., Ebi, K.L., Githeko, A., et al. (Eds.) (2003). *Climate Change and Human Health: Risks and Responses*. Geneva: World Health Organization.

Menne, E., and Ebi, K. (Eds.) (2006). *Climate Change and Adaptation Strategies for Human Health*. Geneva: World Health Organization.

NRSC, ISRO and Department of Ecology, Environment and Remote Sensing, Government of Jammu & Kashmir (2014) A Satellite based Rapid Assessment on Floods in Jammu & Kashmir – September, Table No.6, p. 54.

Raza, M., Ahmad, A., and Mohammad, A. (1978*). The Valley of Kashmir: A Geographical Interpretation* (Vol. 1, pp. 1–59). New Delhi: The Land, Vikas Publishing House Pvt, Ltd..

Stern, N. (2006). *The Economics of Climate Change: The Stern Review*. New York: Cambridge University Press.

Walter Roper Lawrence (1895) *The Valley of Kashmir*. London

14 Perception of Urban Health Hazards in Srinagar, India

Rais Akhtar

Introduction

Man's concern about his environment stems from the threat that environmental pollution poses to his comfort, health and existence (Meade, Florin and Gestler, 1988). Increased awareness and concern are not limited to the problems of air and water pollution discussion of problems of the urban environment, of outdoor recreation, and of threatened animal species appeared with increasing frequency in newspapers and magazines and on radio and television. The discussion has extended to all the problems, present and anticipated that result from rapid growth of population, urbanization, industrialization, and affluence (Saarinen, 1976). In developed countries, like the United States, public opinion polls were conducted in order to measure people's attitudes towards environment one such study carried out in the United States revealed that "the strongest single factor for predicting environmental concern would appear to be educational level".

The more highly educated express greater environmental concern. The less educated and those of lower socio-economic status have the highest proportion who are not very concerned. This lower level of concern among the less affluent may result from their focus on other more pressing problems they must deal with each day (Saarinen, 1976). A number of papers have appeared concerning the geographical perspective of health and health care in urban area. The major emphasis in such studies has been placed on the spatial patterning of ill-health and mortality, physical and human environmental correlates and the organization of health care. However, there exists hardly any work on the understanding of people's perception of health hazards in urban area. Some geographical works which are partially related to the theme includes the work of A. Desai and P. P. Karan, A. Desai focuses her attention on the measurement of perception and its impact on human behaviour in different socio-economic communities in Ahmedabad, India (Desai, 1981). P.P. Karan has made an attempt to measure the people's awareness in his study carried out in Calcutta, India (Karan, 1980, 1981). The author of this paper carried out a study in Lusaka, Zambia, in 1984 with the objective of assessing the variation in perception of urban health hazards in two socio-economically different localities (Akhtar, 1988).

DOI: 10.4324/9781003329459-14

Objectives

The main objective of the present study conducted in Srinagar, India is to focus on the assessment of urban health hazards by people. From the viewpoint of neighbourhood planning, it is desirable to know the differential perception of communities living in socio-economically different residential areas. Their awareness about environmental problems and their willingness to participate in the development programmes can go a long way in improving the health conditions of people living in urban areas.

Methodology

Perception is individualistic and changeable. However, if residents identify themselves with a group and share common beliefs and values, there is a possibility of the emergence of a general pattern in perception and behaviour (Desai, 1981). Data have been collected through a questionnaire pertaining to the subjective images of residents and objective images of investigators with the environmental quality. The environmental quality of each selected (stratified random) household and its surrounding area is evaluated with a score on a five-point scale (i.e., very poor, poor, fair, good, very good) by residents and investigators. This assessment of environmental quality is based on pre-determined variables, including the general appearance and cleanliness of the area, crowding, air pollution, traffic density, housing conditions and presence of mosquitoes and flies. The study is based on 98 households selected from three socio-economically different areas of Srinagar. An assessment of environmental health hazard has been made *by* computing response percentage on a five-point scale for each variable. Besides this, the level of concern pertaining to the environmental quality has also been assessed on a five-point scale (deeply concerned, concerned, somewhat concerned, not concerned and no opinion). Response percentage to total for each of the eight indicators have been computed.

Urban Ecology of the Study Area

The city of Srinagar (M.C.) covers an area of about 177.25 sq. km with a population of 570,195 giving a density of about 3,555 persons per sq.km. The land use pattern of the city shows that the built-up area includes 78.69 sq. km. (44.37%) while the remaining non-built-up areas are covered by the lakes, rivers, nalas, marshy lands, orchards, agricultural lands and graveyards and so on. Thus the built-up area remains very limited while the pressure of population on limited land grows year after year.

The main physiographic characteristics of the site and situation of Srinagar are such that the population is heavily concentrated along both the sides of the river Jhelum and in the adjoining areas. Because of the paucity of residential lands, the population of the city spills over in the low-lying areas, on the slopes of the ridges (ShankaraCharya, the ItariParvat and the Zansker range facing the Dal lake), as well as on the surface of the water bodies (The Dal Lake, the bank of the river Jhelum and the nalas). The residential houses situated in these areas are faced by a number of problems, e.g., drinking water, drainage, sewerage and so on. These problems become more acute during the winter months.

The water-borne diseases are thus more common among the *hanjis* who lives in houseboats and also among the dwellers of the slopping areas where potable water is not available.

The population which is heavily concentrated along the river Jhelum forms compact residential areas. These areas are called the "core areas" of the city. These areas which bear the characteristics of zig-zag roads, narrow lanes, by lanes, dilapidated houses, vertical structures, drainage and sewerage problems, inadequate public amenities and facilities – are some of the major problems which affect the health of the population.

The formation of the core areas is not new to Srinagar alone; it developed in almost all the cities of the medieval period In the recent past, a large scale of exodus from the rural areas has further aggravated the situation over the already congested dwelling so much so that the core areas have become the plethora of urban problems: overcrowding slums, drainage, lack of civic amenities and facilities. Because of the problem of overcrowding, a number of families have started shifting to the outskirts of the city recently. Demographically the areas are over-populated by the fact that an area of 5 sq.km accounts for more than 50% of the total population of Srinagar. According to our estimate, there is a density of 44,000 per sq. kilometre in the core areas including, Rainawari locality, as compared to 1,910 in other parts of the city. This high density can be explained from a cultural perspective. There is a strong tradition of living together in a joint family, and children are greatly attached to their parents. There are rare exceptions where children live separately after their marriage. This has led to overcrowding in the families.

Analysis of Data

Hazratbal

Hazratbal locality is situated along die north-western side of the famous Dal Lake. The area is famous for its Muslim Shrine. An urban health centre and a girl's high school are located in the area.

The movement on the roadside, noise pollution, and traffic congestion have emerged as major health hazards as assessed by the residents (Table 14.1). Since no industry is located in this area, air-pollution is not a problem in Hazratbal. The assessment by the investigator is to some extent also similar to that of the residents (Table 14.2). The investigator has assessed as 'poor' most of the health hazards as well as the availability of amenities in the locality. Table 14.3 reveals the level of concern pertaining to different health hazards. Nearly 67% of residents are 'deeply concerned' about cleanliness/sanitation. Generally people of the locality are either illiterate or less educated, hence they are unable to understand the danger posed by mosquitoes/flies as health hazard.

Rainawari

Rainawari is a high-density area. The majority of the population is self-employed. One hospital as well as one high school and three primary schools are located in the area.

Residents of the area assessed unhygienic conditions as a major health hazard, though at a lesser degree. With the exception of cleanliness-sanitation and traffic density, no other health hazard and the provision of amenities fall under the category of 'very poor' (Table 14.4). The investigator's assessment (Table 14.5) put cleanliness and sanitation under the 'poor' category and the allotted percentage as high as around 81. Cleanliness-sanitation and traffic density have been assessed as 'very poor' by the investigator. As regards the provision of amenities, both school and health facilities are fairly distributed in the locality. This is similar to the assessment made by the residents (Table 14.6). About residents' concern regarding various health hazards, nearly 81% are deeply concerned about the insanitary condition in the area. About 76 and 79% are also concerned about the poor housing condition as well as mosquitoes/flies in the area.

Table 14.1 Assessment by residents (Hazratbal) (% to total for a particular indicator)

	1	*2*	*3*	*4*	*5*	*6*	*7*	*8*	*9*	*10*	*11*	*12*
Very Poor	4.48	15.22	17.40	–	19.56	6.52	–	2.17	2.17	–	–	–
Poor	52.17	71.74	63.04	–	23.91	13.04	67.38	34.78	50.00	58.69	43.48	91.30
Fair	4.35	10.8	8.70	–	10.83	10.87	15.22	28.25	30.44	30.43	50.00	6.53
Good	–	2.17	4.35	–	26.10	39.13	15.22	26.10	10.87	8.71	4.35	2.17
Very Good	–	–	6.51	–	19.60	30.44	2.18	8.70	6.52	2.17	2.17	–

Source: Based on Field Survey.

Table 14.2 Assessment by investigation (Hazratbal) (% to total for a particular indicator)

	1	*2*	*3*	*4*	*5*	*6*	*7*	*8*	*9*	*10*	*11*	*12*
Very Poor	47.83	23.91	15.22	-	6.52	–	2.17	2.17	2.17	–	–	–
Poor	52.17	65.22	58.69	6.52	21.74	43.48	56.54	54.34	60.87	63.04	78.26	73.91
Fair	–	6.53	4.35	2.17	15.22	15.22	30.43	32.61	30.43	30.43	15.21	17.39
Good	–	2.17	8.69	4.35	28.26	15.22	8.69	6.52	6.53	4.36	4.36	2.17
Very Good	–	2.17	13008	86.96	28.26	26.08	2.17	4.36	–	2.17	2.17	6.53

Source: Based on Field Survey.

Table 14.3 To what extent are residents concerned about health hazards (Hazratbal) (% to total for a particular hazard)

	1	*2*	*3*	*4*	*5*	*6*	*7*	*8*
Deeply concerned	67.40	32.61	28.26	32.61	26.29	43.48	17.39	8.69
Concerned	23.91	28.26	39.15	19.56	19.39	32.61	39.13	32.62
Somewhat concerned	6.52	-	-	-	-	-	-	2.17
Not concerned	21.74	10.87	26.09	45.63	-	2.17	-	-
No opinion	2.17	17.39	21.72	21.74	8.69	23.91	41.31	56.52

Source: Based on Field Survey.

Table 14.4 Assessment by residents (Rainawari) (% to total for a particular indicator)

	1	2	3	4	5	6	7	8	9	10	11	12
Very Poor	21.42	–	2.38	–	–	–	–	–	–	–	–	–
Poor	52.00	52.38	47.62	–	35.72	23.81	33.33	28.57	28.57	28.57	28.57	28.57
Fair	7.14	35.72	40.48	–	30.95	23.81	40.48	57.15	57.15	52.42	57.15	52.42
Good	21.44	11.90	9.52	–	30.95	47.62	26.19	14.28	14.28	19.01	14.28	19.01
Very Good	–	–	–	–	2.38	4.76	–	–	–	–	–	–

Source: Based on Field Survey.

Table 14.5 Assessment by investigator (Srinagar Rainawari) (% to total for a particular indicator)

	1	2	3	4	5	6	7	8	9	10	11	12
Very Poor	9.52	–	–	–	2.38	–	–	–	–	–	–	–
Poor	80.96	28.57	30.96	–	19.05	19.05	21.43	23.81	28.57	23.81	23.81	23.81
Fair	4.76	47.62	50.00	–	38.09	28.57	47.62	54.76	54.76	57.14	57.14	57.14
Good	2.38	19.05	14.28	–	33.34	52.38	28.57	21.43	16.67	19.05	19.05	19.05
Very Good	2.38	4.76	4.76	–	7.14	–	2.38	–	–	–	–	–

Source: Based on Field Survey.

Jawahar Nagar

Jawahar Nagar is an upper middle class residential area. The majority of the population is highly educated and work for the government or private institutions. According to the assessment by residents (Table 14.7) cleanliness – sanitation is not a problem in the area. However, the residents consider noise and air pollution as hazards in the area. This is because of the fact that most people own their cars, as well as impact of air pollution.

Amenities such as transport, health and school are sufficiently available in the area. The roads are quite wide and thus no problem regarding the movement on the roadside. The investigator has not assessed any hazard and the provision of amenities under 'very poor' category (Table 14.8). Similar to the assessment made by residents, the investigator has also assessed hazard related to mosquito/flies as 'poor'. The assessment made by investigator shows that water pollution is a problem and categorised it as 'poor'. This is probably the impact of the pollutants from the industries. Amenities such as transport, health as well as movement on the roadside have been assessed as 'fair' by the investigator. Nearly 70% of residents are "deeply concerned" about the cleanliness-sanitation, while only 30% are 'somewhat concerned' about it (Table 14.9). Contrary to this only about 20% each are concerned with the problems related to traffic density and air pollution. About 70 and 60% are 'somewhat concerned' regarding the housing condition and mosquitoes/flies respectively.

Table 14.6 To what extent are residents concerned about health hazards (Rainawari) (% to total for a particular hazard)

	1	2	3	4	5	6	7	8
Deeply concerned	80.96	33.34	54.76	–	–	38.09	4.76	2.38
Concerned	16.66	54.76	38.10	–	42.86	76.19	78.57	–
Somewhat concerned	2.38	4.76	-	–	–	7.14	2.38	–
Not concerned	–	7.14	7.14	–	–	2.39	–	–
No opinion	–	–	–	–	–	9.52	16.67	19.05

Source: Based on Field Survey.

Table 14.7 Assessment by residents (Jawahar Nagar) (% to total for a particular indicator)

	1	2	3	4	5	6	7	8
Very poor	–	10	10	60	50	50	–	20
Poor	–	70	70	30	40	50	–	80
Fair	60	20	20	10	10	–	10	–
Good	38	–	–	–	–	–	70	–
Very good	10	–	–	–	–	–	20	–

Source: Based on Field Survey.

Table 14.8 Assessment by investigator (Jawahar Nagar) (% to total for a particular indicator)

	1	2	3	4	5	6	7	8
Very poor	–	–	–	–	30	–	–	–
Poor	–	70	60	80	40	90	–	80
Fair	40	30	40	20	30	10	30	20
Good	50	–	–	–	–	–	50	–
Very good	10	–	–	–	–	–	20	–

Source: Based on Field Survey.

Conclusion

The study was aimed at determining the seriousness of urban health hazards as perceived by the residents. The study reveals that range of individual perception of environmental quality varies to some extent in different areas. However, there are no significant variations between the assessment by the residents and the investigator. Also evident is the fact that residents assessed their environment not only as individuals but as an entire community. The study also highlights the fact that the variations in the assessment of health hazards is the relation of the socio-economic structure of the population concerned The economically and educationally well off assessed noise and air pollution as health hazards while they are quite 'deeply concerned' with cleanliness-sanitation. Contrary to this, residents in other medium and high-density areas assessed unhygienic environment and traffic density as a health hazard.

Table 14.9 To what extent residents are concerned about health hazards (Jawahar Nagar) (% to total for a particular hazard)

	1	2	3	4	5	6	7	8
Deeply concerned	70	–	20	20	–	–	30	–
Concerned	–	–	–	–	–	–	–	–
Somewhat concerned	30	50	30	50	–	20	70	60
Not concerned	–	20	50	30	–	40	–	40
No opinion	–	30	–	–	100	40	–	–

Source: Based on Field Survey.

Besides this, the description of data concerning the people's perception, their housing conditions and their locality, the availability of amenities in the area may be useful in the socio-economic health planning of the area concerned. It must be added that the core area development programme being initiated in Srinagar is aimed at the development of parks, removal of slums, widening of roads, repair and replacement of old bridges. These measures if implemented may considerably reduce the unhygienic conditions in urban Srinagar.

Acknowledgement

I am grateful to my post-graduate students and research scholars who assisted me in obtaining data. Also I am thankful to Dr. M Siddiq for his comments on aspects of urban ecology of Srinagar.

References

Akhtar, R. (1988). Perception of Urban Health Hazards: Examples from Lusaka, Zambia, *International Journal of Environmental Studies,* 31, 2-3, pp. 167–172.

Desai, A. (1981). Differential Perception of Residents to Environmental Quality of an Urban Area: The Case of Ahmedabad, *Geographical Review of India,* 43, 2, pp. 156–165, pp. 198–199.

Karan, P.P. (1980–82). Public Awareness of Environmental problems in Calcutta Metropolitan *Area, National Geographical Journal of India,* 26–28, pp. 29–34.

Meade, M., Florin, I. and Gestler, W. (1988).*Medical Geography,* New York, The Guilford Press, p. 167.

Saarinen, T.F. (1976). *Environmental Planning: Perception and Behaviour,* Atlanta, Houghton Mifflin Company, p. 198.

15 Health Care Behaviour in Western Himalayas

A Case Study of Udhampur Forest Villagers in Jammu and Kashmir

Anuradha Sharma, Sandeep Singh, and Davinder Singh

Introduction

Man has long been associated with the plant kingdom since his existence due to their unique medical value, safe effective and inexpensive indigenous remedies that are fast gaining popularity. WHO (World Health Organization) approximates that 80% of people rely on herbal medicines partly for their primary physical condition (Molateb 2011). It also estimates that up to 68,000 plant species are used in folk medicine and a vastness of this variety is found in the Asia-Pacific region. The traditional system of medicine mainly functions through two distinct streams (a) Local or folk or tribal stream and, (b) Codified and organized Indian system of medicines like Ayurveda Siddha and Unanni. Although synthetic drugs are often used in treatment of certain disease but a remarkable interest and confidence on plant medicine has also been found (Tiwari 2008). **Indian Vedas** describe the widespread use of herbal products and aqueous extract of different plant parts for curing different diseases. Thirty percent of root part of medicinal plant is used in different practices in comparison to other plant parts (Ved et al. 1998).

India has been identified as one of the top 12 mega bio-diversity centres of the world due to vast area with wide variation in climate, soil, altitude and latitude. India with its biggest repository of medicinal plants in the world may maintain an important position in the production of raw materials either directly for crude drugs or as the bioactive compounds in the formulation of pharmaceuticals and cosmetics. In India alone 17% out of 15,000 species of flowering plants are of medicinal value. Most of the plants are collected from nature indiscriminately for commercial exploitation, resulting in depletion hence the inventory of medicinal plant species which evolved and adopted over a long period, assumes great significance (Dalal 1998). From time immemorial different plants are used as medicine in our country (Chopra et al. 1956). We have always made use of flora to alleviate sufferings and diseases. The system of ethno-medicine is safe and is a low-cost therapy for treating a variety of ailments. As elsewhere, in India too, the medicinal use of plants has been practiced while ancient times by various rural and tribal community through the system of Ayurveda, Siddha and Unani (Gadgil 1996).

The endemism is largely confined to undisturbed or fragmented natural forests. Many of the endemic species are forest specialists depending on diverse and dense

DOI: 10.4324/9781003329459-15

primary forests at a given altitudinal range, (Azeria et al. 2007) recognizes the rank of endangered medicinal plant species in HP (Himachal Pradesh), focused on two step ex-situ cultivation i.e. short-term and long-term. Four basic criteria were considered to prioritize species, endangered status, knowledge base studies, increased technology, farming prospects and marketability (Badola & Pal 2002). Fourteen species were acknowledged to be cultivated in four agro-climatic zones. Species identified for each zone was assessed for their selected populations, accessibility of propagates and quality planting stock, certification, quality production, value addition and ensured markets. In Kathua, where the major components of collection/consumption of fuelwood and fodder besides medicinal were considered an analysis of caste, income and distance from the road was done and it was observed that these indicators greatly influenced the collection/consumption pattern (Bhushan 2014). In forest resources utilization study of Jammu Region the quantification of fuelwood and fodder at various attitudinal levels indicated that the people living in kandi areas and high altitudes of Jammu are suffering from the inadequacies where the socio-economic setup is in deplorable condition (Dubey 1995). The quality and quantity of nutrients availability, one of the major hindrances in ruminant production in Pakistan where the area under forest is continuously reducing indicating high pressure for cash crops production. The tree leaves supplement the existing feed resources for small as well as large ruminant and can help to bridge the wider gap between demand and supply of nutrients. Tree leaves are a rich source of supplementary protein, vitamins and minerals and their use in ruminant to enhance microbial growth and digestion (Cheema et al. 2011). The ethnobotanical revision of useful shrubs of district Kotli in Azad Kashmir and reported 38 species of 36 genera belonging to 25 families. Most of the shrubs were found useful in everyday life of the local inhabitants, which were being used as medicines, fuel, shelter, and fodder/forage in manufacture agricultural tools. Most of the shrubs were having more than one ethno botanical utilization (Ajaib et al. 2010) the status of medicinal plants consumption by the pharmaceutical industries in Gujarat state, created the baseline database for indigenous and imported such as cultivated trees, shrubs, climbers, herbs and different crops used in Ayurvedic medicines. Gujarat pharmaceutical companies consume all the 310 herbal raw material obtained from 270 plant species. With 148 species reported from forests, the consumption of vital parts like root, bark, seed, gum and whole plant are most detrimental to the ecology in general and the vegetation in particular . These plants and trees are of immanence importance and require a detailed study to make things commercially significant (Singh & Paraiba 2003).

Objective

The western Himalayan region of India is a storehouse of rich medicinal plants where people residing in forests heavily depend on this resource for dealing with their day-to-day requirements including medicinal remedies with great opportunities especially to poor people. This paper attempts to understand the utilization pattern of medicinal plants and their attitude towards health care behaviour of Western Himalayan dwellers.

Methodology

The study focuses on selected background characteristics of the sample respondents, their preference towards the medical facilities besides traditional know how of medicinal plants which are being regularly used in day–to-day life of these Himalayan dwellers. The study area was divided into four zones on the basis of Physioclimatic conditions representing tropical and temperate type of climate.

The zones were identified as tropical, sub-tropical, temperate and sub temperate ranging from 400–1,200 mts, 1,200–2,000 mts, 2,000–2,800 mts and 2,800 mts and above respectively. These four zones were randomly surveyed with the help of questionnaire that was prepared keeping in view the socio-economic parameters selected for the study such as caste, income and distance from the road. In total 400 households were interviewed, selecting 100 each from each zone.

Table 15.1 reflects the name of species occurring in all four zones. Four transitional/tranzitional zones have been identity filed in the study area as mixture of species is available. Many species in the area have been recognized as undamaged.

Table 15.1 Zone-wise species of Udhampur forest

(Botanical name) local name	*Zone*
Katha/khair (Acacia catechu), Kikar (Acacia nilotica), Farlai (Acacia modesta), Bel (Aeglemarmelos), karyar/amaltus (Cassia fistula), Tali/Shisham (Dalbergiasissoo), Thor (Euphorbia royleana), Bar (Ficusbengalensis), Rumbal (Ficusracemosa), Papal (Ficusreligiosa), Gamber (Grewiaarborea), Dhaman (Grewia elastic), Bakli(Lagerstroemia arvifolia), Am (Mangiferaindica), Dhrenk (Meliaazedarach), Tut (Morusseratta)	I
Chiru (Pinuslongifolia), Goon(Aesculusindica), karengal (Bauhinia retusa), Paleh/kinjoo/Tatootye (Buteamonosperma), kharik (Celtisaustralis), Amla (Emblicaofficinalis), Jamun (Syzygiumcuminii)	T.Z -1(I,II)
Karer (Bauhinia variegate), Thum (Erythrinaindica), trembal (Ficusauriculata), phagwara (Ficus palmate), gamber (Gmelinaarborea), chandri (Machilusduthiei), Kalam (Mitragynaparvifolia), Koa (Olea cuspidate), Ailan (Pierisovalifolia), Chill/chir (Pinusroxburghii), Kakar (Pistaciaintegerrima), kalkat (Prunuspadus), Banj (Quercusleucotrichophora), Dadri (Toonaserrata)	II
kharshu (Quercussettecarpifolia) Siris (Albizzialebbek) champ (Alnusnitida) Dheo,bakli (Anogeissuslatifolia) Brenkar (Agave Americana) Simbal (Bonbaxceiba) Swalia (Colebrookiapositifoli) Luni (Cotoneaster icrophyllakharsu) (Quercussemecarpifolia) Karal (Bauhinia purpurea) zhinjera (Bauhinia racemosa) chikri (Buxuxwallichiana)	T.Z -2 (II,III)
Thangi/Badam (Coryluscolurna), Rheu (Cotoneaster bacillaris), deodar (Cedrusdeodara), Akhrot (Juglansregia), Kemba, simla (Lanneacoromandelica), Kamila (Machilusphillippinensis), khajur (Phoenix sylvestris), Byad/kail (Pinnuswallichiana), Lami (Pinuspadus), kainth (Pyruspashia), Moru (Quercusdilatata), Bras/chew (Rhododendron arboretum), Chimber (Rosa macrophylla), Arjan (Terminaliaarjuna), Toon (Toona ciliate), Behra (Viburnum cotinifolium), Lunee (Viburnum fatens)	III

(Continued)

Table 15.1 (Continued)

(Botanical name) local name	Zone
lamon (Daphne cannabina) charmra (Desmodiumiliaefolium) kakrel (Dioscoreaspp) santha (Dodonaea viscose) traporthor (Opuntiadillenii)	I,II,III
Chest tree (Vitexnegundo) Fagara (Zanthoxylumalatum) Karu (Aesculusindica) Raspberry (Rubusellipticus) Barberry (Berberis lyceum) Berries (Daphne cannabina) Deodar (Cedrusdeodara) fir (Abiespindrow)	T.Z -3 (III,IV)
Beek (Quercussermecarpifolia) Chadar (Betulaalnoides) Acer (Acer caesium) Burans (Rhododendron arboretum) Bird Cherry (Prunuspadus) Bistort (Polygonumspp) Wild strawberry (Fragariavasica) Pansy (Viola) valerians (Valerianaspp) Dhoob (Primuta) Phooli (Anemon) Spruce (Piceaengelmannii)	IV

Source: Organized by author (TZ is the transitional zone where mixture of species exist).

Results and Analysis

Detailed account of quantity of medicinal plants used by each household in one month based on different socioeconomic parameters at each zone.

The table clearly indicates that the percentage of users and quantity used increased with altitude. The same trend has been observed in income groups and distance from road. Lesser is the income more is quantity used. It is interesting to note that there is almost four times variation from lower to higher income group. As the distance from the road increased the quantity of medicinal plants also doubled.

The average quantity of medicinal plants used by each household is given in Table 15.2 where in zone I Most of the general caste households are extracting these plants for selling them into the market and earn their livelihood where 90% of general and 91% of ST population is involved. In zone II 95% of the households belonging to general caste use 4.5 kg/month/hh, in their day to day life. In this zone only 80% households belonging to OBC use medicinal plants in their daily routine with an average quantity of 1.25 kg/month/hh, 90% of the SC and ST households use these plants with an average of 1.5 kg/month/hh and 2.5 kg/month/hh respectively. Most of the OBC and SC households are engaged in the labour work, so their percentage usage is lower than the other caste.

The income wise data reflects that with increase in income the usage decreases. Households belonging to lower income groups is above 90% but their monthly usage varies as it is 3.7 kg/month/hh in case of lower income group and 1.7 kg/month/hh in case of second income group. The percentage of two income groups decreases i.e. 90% of the third income group i.e. between Rs. 25,000 to Rs. 40,000 use at an average of 1.0 kg/month/hh and 85% belonging to the higher income group used 1.0 kg/month/hh.

The above table also reflects the percentage of households using medicinal plants as per the distance from the roadside. It reflects the same pattern as in zone II with increase in distance the usage also increases. It shows that 89% of the households living along the roads or nearby settlements, use 1.00 kg/month/hh, but

Table 15.2 Average quantity of households using medicinal plants in study area

Caste zone	Percentage I II III IV	Quantity used (in kg/month/ hh) I II III IV
General	90 95 97 100	1.8 4.5 3.5 6.5
OBC	79 80 91 100	2.1 1.2 1.7 2.9
SC	86 90 93 100	1.9 1.5 1.5 2.3
ST	91 90 95 100	2.4 2.5 1.2 2.4
Income (in Rs/month)		
Less than 10,000	95 96 100 100	2.5 3.7 3.25 7.3
10,000-25,000	86 91 95 100	2.1 1.7 1.95 4.0
25,000-40,000	77 90 90 100	1.5 1.0 1.50 2.1
More than 40,000	65 85 90 100	1.2 1.0 1.00 1.0
Distance (in km)		
On roadside	76 89 92 100	1.7 1.0 1.2 2.1
Between 1 and 4	82 94 94 100	2.0 2.0 2.0 5.0
Between 4 and 8	91 96 100 100	2.5 2.0 3.1 7.1

Source: Field survey, 2016.

as the distance increases the percentage of households using medicinal plants also increases to 94% and 96% for the households living between 1–4kms and 4–8kms respectively but their quantity used remains the same, i.e. 2.0 kg/month/hh in zone II and zone III but as the distance increases in the percentage of households the quantity used also increases significantly. This pattern is seen in all the four zones.

There are number of medicinal plants found in zone III such as Kakoa, Datura, Seski, Arjan, etc. All these plants are of medicinal as well as economic importance. The general caste households having income less than Rs. 10,000 are engaged in this activity to earn their livelihood. The table reveals that 97% of the General caste households use 3.5 kg/month. In this zone 91% households belonging to OBC caste use an average quantity of 1.7 kg/month/hh, 93% of the SC households and 95% of ST households use an average of 1.5 kg/month/hh and 1.2 kg/month/hh respectively.

The income wise percentage of households again reflects the same pattern as in lower zones that as the income of the households increase, the percentage of households using medicinal plants decreased. It is noticed that almost every household use medicinal plants but the households belonging to lower income group are using this valuable resource for their livelihood. Every household belonging to lower income group, i.e. less than Rs. 10,000/month used 3.2 kg. In case of second-income group, i.e. between Rs. 10,000/month and 25,000/month the percentage of households using medicinal plants above 95% and their monthly usage is 1.9 kg/month/hh. Some households belonging to these income groups use these plants and sell them in the market too. The percentage of household in the other two income groups is 90%, but their quantity used varies, i.e. the households belonging to the third-income group between Rs. 25,000/month and Rs. 40,000/month use medicinal plants with an average of 1.5 kg/month/hh and the households belonging to the higher-income group (more than Rs. 40,000) used 1.0 kg/month/hh.

In zone III 92% households living along the roads or nearby settlements, use 1.2 kg/month/hh, but this increases to 94% and 100% for the households living between 1–4 kms and 4–8kms respectively along with their quantity, i.e. 2.0 kg/month/hh and 3.1 kg/month/hh as in case of other zones.

The medicinal plants available in zone IV are Dhup, Banafsa, Kaur, Brahmibooti, etc. All these plants are of medicinal as well as economic importance. In this zone, lower-income group is engaged in this activity to earn their livelihood. The table reveals that 100% of the households found in this zone are using these plants in their daily routine, but the percentage varies with respect to caste and it is only due to the engagement of households in extraction activity. The households belonging to general caste use 6.5 kg/month/hh, it is 2.95 kg/month/hh in case of OBC, 2.3 kg/month/hh in case of SC and 2.45 kg/month/hh in case of ST households.

In this zone all the income groups use these plants on day to day basis and variation in their usage has been recorded in lowest income group that uses 7.3 kg/month/hh followed by increasing income groups with 4.0, 2.1 and 1.00 kg/month/hh respectively.

In this zone the road connectivity is not good as in the other zones. It was observed that every household in the zone uses medicinal plants and as the distance from the road increases deep into the forests, the quantity of using medicinal plants increases. It is 2.1 kg/month/hh for the households living along the roads or nearby settlements and for the households living between 1–4 kms and 4–8 kms it is 5.0 kg/month/hh and 7.1 kg/month/hh respectively (Table 15.3).

The above table reveals that the traditional treatment is very popular followed by allopathic and self-medication. Individually speaking in zone-I and zone-II allopathic treatment is preferred whereas in higher zones the tradition treatment is opted.

The overall population in the health care behaviour of high altitude region reflected that at an average only 13.7 follow self-medication, 6.7% discuss their problem with friends/relatives, and only 23% go for allopathic treatment, The overall picture reveals that in study area the highest average 37% is that of traditional treatment followed by allopathy 35.5%, self-medication 21.4% and only 5.7% at an average discuss their problems with friends and relatives. In this zone no Vaid/Hakim was available hence people only relied on the family traditional knowhow

Table 15.3 Health care behaviour

Zone	Buy medicine from shops (self-medication)	Discuss with friends	Allopathic treatment	Traditional treatment
Zone I	28	4	44.7	23.2
Zone II	24.7	5.7	39.5	30
Zone III	19.2	6.5	34.7	39.5
Zone IV	13.7	6.7	23	56.5
Average	21.4	5.7	35.5	37.3

Source: Field survey, 2016.

Table 15.4 Disease-specific health care behaviour

Zone	Buy medicine from shops (self-medication)				Discuss with friends				Allopathic treatment				Traditional treatment			
	D	C	E	W	D	C	E	W	D	C	E	W	D	C	E	W
ZONE 1	22	26	43	21	5	7	3	1	45	42	32	60	28	25	22	18
ZONE II	20	25	35	19	6	9	6	2	35	37	34	52	39	29	25	27
ZONE III	15	17	30	15	10	5	7	4	23	28	37	51	52	50	26	30
ZONE IV	6	5	8	8	3	3	4	3	9	8	13	16	32	34	25	23

Source: Field survey, 2016.
Note: Diarrhoea = D, Cold/Cough = C, Eye diseases = E, Wounds/Cut = W.

for the ailment. The respondent of mid two zones also showed their inclination towards the traditional treatment and in the lower altitude the Allopathic treatment was found to be more popular (44.7%).

An attempt has been made to understand the health care behaviour of the people of the study area for common diseases. The common diseases consider for the study were diarrhoea, cold/cough, eye disease and cuts/wounds. It was also tried to understand their preferences for traditional and allopathic treatment besides their behaviour of the dealing with the common diseases as whether they buy the medicine from the chemist shop by their own knowledge or discuss the problems with their friend/relative to seek their suggestion. The medicine they buy the shop is without the consultation of any specialist.

Table 15.4 gives an overall behaviour of the study area. Interestingly in zone I, to treat diarrhoea 28% buy the medicine from the shop without consultation and the decrease has been recorded in following zones as we move to zone IV with only 6%. The same trend appears for other diseases too, which decrease as we move towards higher altitude sharing and seeking the advice from the friends/relatives seems more common in zone III where the percentage is little higher than the rest. The allopathic treatment for all the diseases gain popularity in the lower altitudes than higher except for eye diseases in zone II and zone III where a little change has been recorded. Interestingly the traditional treatment for all the diseases is more commonly preferred as we move higher but zone IV records lesser popularity of traditional treatment.

Conclusion

Udhampur forest are rich in medicinal plants and people use them for treating their day-to-day problems with their traditional knowledge. In the lower zones, though percentage of households and the quantity extracted is less, still more than 80% are using, these resources irrespective of their caste and location.

Variation has been observed with respect to income where in, below 1200 mts 65% households are using locally available plants but as we go higher, the percentage increases subsequently making it 100% in zone IV. This clearly indicates that people of western Himalayas have unconditional dependency on forest resources for their wellness and health which is clearly evident from their behavioural pattern.

References:

Ajaib, M., Khan, Z., Khan, N., & Wahab, M. (2010). Ethnobotanical Studies on useful Shrubs of District Kotli, Azad Jammu & Kashmir, and Pakistan. *Pakistan Journal of Botany, 42*(3), 1407–1415.

Azeria et al. (2007). Biogeographic patterns of the East African coastal forest vertebrate fauna. *Biodiversity and Conservation, 16*(4), 883–912.

Badola, H.K., & Pal, M. (2002). Endangered medicinal plant species in Himachal Pradesh. *Current Science, 83*(7), 797–798.

Bhushan, I. (2014). Evaluation of forest resources utilization in kathua forest division: A study in resource geography. Unpublished Ph.D thesis, Department of Geography, University of Jammu-Jammu

Cheema, U.B., Younis, M., & Sultan, J.I. (2011). Fodder tree leaves: an alternative source of livestock feeding, Advances *in Agricultural Biotechnology, 2*, 22–33.

Chopra, R.N., Nayar, S.L., & Chopra, I.C., (1956). Glossary of India Medicinal Plants, CSIR, New Delhi.

Dalal, P. (1998). Strategy for conservation and availability of Medicinal Plants. Abstracts of Nat. Symp. On species, Medicinal and Aromatic Plants-Biodiversity, conservation and utilization (NSMAP), Calicut, Kerala, p. 14.

Dubey, A. (1995). Some aspects of the forest resources utilization of Jammu Region: A spatio-temporal focus. Unpublished Ph.D thesis, Department of Geography, University of Jammu-Jammu.

Gadgil, M. (1996). Documenting diversity; an experiment. *Current Science, 70*(1), 36.

Molateb, M.A. (2011). Selected medicinal plants of Chittagong hill tracts. *International Union for Conservation of Nature,* Dhaka, Bangladesh, pp xii + 116.

Singh, A.P., & Parabia, M. (2003). Status of medicinal plants consumption by pharmaceutical industries in Gujarat state. *Indian Forester, 129*(2), 198–212.

Tiwari, S. (2008). A Rich Source of Herbal Medicine. *Journal of Natural Products, 1*, 27–35.

Ved, D.K., Mudappa, A., & Shankar, D. (1998). Regulating export of endangered medicinal plant species-need for scientific vigour. *Current Science, 75*, 341–344.

16 Health-seeking Behaviour among the Hanjis of Dal Lake, Kashmir, India

Mushtaq A. Kumar, Shazia Mehraj and Dr. G.M. Rather

Introduction

Health is not merely the absence of disease and infirmity, it is a state of total physical, mental and social well-being (Ryff & Singer, 1998). Health is a broad concept that embraces all social and biological aspects of life, and health seeking refers to a series of acts made to maintain or prevent health (WHO, 1997). The steps of self-evaluation of systems, self-treatment, seeking professional help, and acting on expert advice are all part of the dynamic process of seeking healthcare behaviour (Gupta, 2010). Health-seeking behaviour is a common practice among members of a community that arises from the interaction and balance of health requirements, health resources, and socioeconomic, cultural, political, and national/ international contextual factors (Adhikari& Rijal, 2014). Individuals' attempts to rectify perceived dangers to their health and well-being are referred to as health-seeking behaviour (Oberoi et al., 2016).

Healthcare-seeking behaviour is defined as decision-making for healthcare at the household level wherein the decisions made encompass all available options: public, private, modern and traditional. Health-seeking behaviour has been defined as any action or inaction undertaken by individuals who perceive themselves to have health problems or to be ill to find an appropriate remedy (Latunji & Akinyemi, 2018). Health-seeking behaviour is situated within the broader concept of health behaviour, which encompasses activities undertaken to maintain good health, to prevent ill health as well as dealing with any departure from a good state of health. In other words, health-seeking behaviour is the habit of people concerning their health. This behaviour consists of inattentive to illness, treatment action that means people visit health providers, and self-medication (Dawood et al., 2017). The country's healthcare system and the health service have a direct impact on healthcare utilization (Abuduxike et al., 2020; Andersen & Newman, 1973). The act of seeking medical attention has enormous potential for lowering the incidence of disease, disability and mortality (WHO, 1997). The main driver for health-seeking behaviour is the organization of the healthcare system. There can be two ways to understand health-seeking behaviour. First understanding health-seeking behaviour by emphasizing the 'end point' which is the utilization of healthcare systems or healthcare-seeking behaviour. Second, by stressing the 'process,' which

DOI: 10.4324/9781003329459-16

refers to the activities that lead to an illness reaction or health-seeking behaviour. Appropriate health-seeking behaviour is critical for reducing complications and improving life quality (Abidin et al., 2014).

Healthcare behaviour is a complex outcome of many factors operating at individual, family and community levels (Vaishnavi & Mishra, 2021). Treatment choices would involve many factors related to illness causation, accessibility of treatment available and their perceived efficacy and disease profile (Inchi et al., 2014). According to sociology literature, healthcare-seeking behaviour is influenced by the individual self, diseases and the availability and accessibility of health services. Individuals differ in their sources, treatment of choice (Kakai et al., 2003). The concept of studying health-seeking behaviour has evolved with time. These days, it has become a tool for understanding how people engage with healthcare systems in their respective socio-cultural, economic and demographic circumstances. Individuals differ in their willingness to seek medical assistance. Some people seek treatment without hesitation, while others wait until they have reached the advanced level of ill health. Healthcare-seeking behaviour is influenced by availability, quality and price of services as well as to social groups, health view, residences and personal features of a user (Begashaw et al., 2016). People's healthcare behaviour is influenced by their access to healthcare facilities, as well as the cost of treatment and the attitude of healthcare providers. According to Bojola et al. (2018), health services and resources remain inadequately used. Furthermore, people's healthcare preferences vary based on socio-demographic, socio-economic and cultural factors, all of which influence their healthcare-seeking behaviour. People who live in urban areas are thought to be more open to new ideas and eager to attempt new things on a trial-and-error basis. On the other hand, rural residents are tradition-bound, resistant to change, and eager to hold on to traditional beliefs and customs.

Shreds of evidence show that socio-economic status, geographic settings, cultural issues, service quality, health system policy and procedures are among the factors affecting health-seeking behaviour of the community (Hoeven et al., 2012; Iyalomhe & Iyalomhe, 2012). Individuals who fail to get health information are found to have lower health-seeking behaviour (Rahman et al., 2011). However, individuals having higher health-seeking behaviour could better prevent disease and promote health (Iyalomhe & Iyalomhe, 2012; Macro, 2006). Healthcare-seeking behaviour is a multifaceted effect and needs an appropriate investigation to provide knowledge that will help the formulation of healthcare policies and programs (Asfaw et al., 2018).

Planning health-care policies and programs necessitate knowledge of healthcare seeking in order to ensure early diagnosis, successful treatment, and the implementation of relevant interventions. The practice of healthcare-seeking behaviour has a marvellous potential to reduce morbidity, disability and mortality. Understanding healthcare-seeking behaviour and its causes help the government, stakeholders, policymakers and healthcare professionals in allocating and managing available resources effectively. To combat expensive healthcare prices, it is critical to understand the healthcare-seeking habits of various communities and population groups.

Healthcare-seeking behaviour is closely linked with the health status of a nation and thus its economic development. The purpose of this study was to look into the healthcare behaviours of the Hanjis of Dal Lake in terms of disease perception and medical system utilization.

Study Area

Dal is lake in Srinagar, the summer capital of Jammu and Kashmir. Dal lake lies to the east and north of the Srinagar city, located between 34°04′–34°11′ N latitude and 74°48′–74°53′ E longitude (Ali, 2015). The length of lake is 7.44 km (2.2 miles). The average elevation of lake is 1,583 m (5,190 ft). The lake has a total area of above 22 km² of which approximately 12 km² is the total open water spread area. The total area of its catchment is about 316 km². The shore line of the lake is about 15.5 kilometers. The shallow lake is fed by Dachigam Telbal nallah, Dara Nallah and many other streams. The lake is in the foothill formations of the catchment of the Zabarwan mountain valley, in the foothills of the Shankaracharaya hill which surround it three sides (Bhat et al., 2017) It is integral to tourism and recreation in Kashmir and is named as the "lake of flowers", "Jewel in the crown of Kashmir" or "Srinagar's Jewel". It has been epitome of the Kashmiri civilization and has played a major role in the economy of the state through its attraction of tourists as well as its utilization as a source of food and water. A large number of gardens and orchards have been laid along the shores. The lake is famous not only for its beauty, but also for its vibrancy, because it sustains within its periphery, a life that is unique anywhere in the world.

Database and Methodology

The survey is based on primary sources of data. The primary information was drawn from households from different areas of Dal Lake through stratified random sampling technique. A total of 112 household samples were collected from five different Mohallas of Dal Lake viz Shabri Mohalla, Kanketi, Sheikh Mohalla, Mir Mohalla and Kabutar Khana.

A well-developed questionnaire was used to obtain data from the participants. The questionnaire composed of four sections mainly to evaluate the healthcare behaviour among the Hanji community. The first section obtained the sociodemographic data of respondents including age, gender, education level, occupation, marital status and family size. The second section included status of respondent's health, presence of chronic disease, utilization of health services such as availability of type of facility, preferred medical systems, reasons to choose a particular facility, accessibility of health facility. The third section included perception about disease such as causes, control over disease and understanding of health problem. The fourth part included management of health problems such as undertake investigations, undertake health-oriented activities, discuss problem with others, etc.

Data tabulation: whole data was tabulated systematically in Microsoft excel. The results were drawn with the help of simple mathematical techniques.

Socio-Economic Profile of Hanjis

The segment of population living in boats of different shapes, sizes and types and earn their livelihoods in and around the different water-bodies especially in Dal Lake (Fazal and Amin, 2012). The boatman of Kashmir is known as Ha'enz in local language and Hanji in Hindi script. They are also called as Kishtiban (boatman) or Jalbashi (water dwellers). Hanjis are among the aboriginal inhabitants of Kashmir valley and are prominent ethnic community of the valley. Hanjis draw their livelihood from the lake and carrying out activities such as; water transporters, fishermen, vegetables-growers, wood cutters, grain carriers, dealers of construction materials, collectors of various lake products, paying guest keepers and tourist guide. Hanjis have been identified as a separate caste in the union territory of Jammu and Kashmir. Hanjis live in and around the Dal Lake in different localities which are commonly known as Mohallas. The form, structure and size of these Mohallas vary from one locality to another – some are big consisting hundreds of households while some are small. The Dalgate and Gagribal of the lake have higher concentration of Hanjis. In the Dal Lake, Hanjis cover an area of sixteen (16) localities (villages). The Hanji localities although disperse in their location within and around Dal Lake but they connect themselves mainly by the water ways and rarely by the land routes. Hanjis are divisible into different types according to their social status and occupation. On the basis of different economic activities and social setup Hanjis can be categorised into following categories:

> Demb Haenz (vegetable growers), Gari Haenz (water-nut gatherers), Dung Haenz (owners of passenger boats, dunga), Ma'er Haenz (boatman of Ma'er Nallah), Gaad Haenz (fisherman), Haka Haenz (collectors of wood from water bodies), Shikara Haenz (shikara owners), Houseboat Haenz (houseboat owners), Bahatchi Haenz (who live in Bahatch boats). Economically, the hanjis are backward. They do not have fixed income and their income varies from month to month and season to season. Among the different types of Hanjis, Demb hanjis and Gaad hanjis are the poorest. The owners of houseboat, however are rich, maintaining a good standard of living. Houseboat area of Dal Lake differs from the hamlet population due to its proximity to the city and interaction with the tourists. Houseboat owners are not generally into cultivation though they are found to own some water areas. Very few hanjis in the dal lake are literate. The literacy rate among the Hanjis of Dal Lake is poor. According to one field study literacy rate among Hanjis is only about 24%. Among the Hanjis the male and female literacy rate are 21 and 5% respectively. The Dal Lake is the abode of Hanjis and provides them sustenance. The Demb, Gaad and Shikara Hanjis are almost exclusively dependent on it. Their mode of life has been closely influenced by the resource potential of the lake.

Table 16.1 shows the socio-demographic profile of respondents. A total of 112 households were surveyed. As per the results, 66.07% respondents were females

Table 16.1 Socio-demographic profile of respondents

Variables	Percentage
Gender	
Male	33.92
Female	66.07
Age	
0–14	12.20
15–59	79.20
60 and above	8.60
Education	
Illiterate	71.42
Elementary	5.35
Intermediate	8.92
High school	5.53
Graduate	8.92
Marital status	
Married	73.21
Unmarried	26.79
Monthly income	
> 15,000	55.35
< 15,000	44.65
Type of family	
Nuclear	75.00
Joint	25.00
Family size	
< 5	42.85
> 5	57.14

Source: Based on Data obtained from Field work 2021

and 33.92% were males. The sample distribution by age reveals that 12.20%, 79.20% and 8.60% of the respondents fall in the age group of 0–14, 15–59 and above 60 years respectively. Furthermore, majority of respondents were illiterate (71.42%), followed by 8.92% respondents who have studied up to secondary, 8.92% were under graduates and 5.35% of respondents have studied up to higher secondary and 5.35% had studied up to primary levels. Out of the total respondents 73.21% were married and 26.79% were unmarried. In case of monthly income, 55.35% respondents earned less than 15,000 per month, and 44.65% of respondents who earned more than 15,000 per month. Out of the total study subjects, 75% of respondents had nuclear families and 25% of respondents had joint families. So, for as the family size is concerned, 57.14% of respondents had five or more than members in their family while as 42.85% had less than five members in their family.

Table 16.2 Perception of respondents about their own health

Variable (%age)	Poor (1)	Average (2)	Good (3)	Better (4)
Health status	21.42	23.21	41.07	14.28

Source: Based on Data obtained from Field work 2021.

Perception of Health

When asked to rate their current health status (Table 16.2), 41.07% of respondents perceived their current health status as good, 23.2% of the respondents as average, 21.42% as poor and 14.28% as better.

Utilization of Medical System

Table 16.3 depicts that 90.16% of respondents preferred allopathic health system while as 3.57% of the respondents preferred traditional healers, while as 6.27% of respondents chose homeopathy or Unani system. 55.35% of respondents selected private medical facility as medical institution of first choice and 44.64% selected public medical facility. The reason for choosing a particular facility were, 71.42% of respondents chose it for quality of treatment, 21.24% respondents chose it as it near to home and 7.14% for affordability. 58.92% of respondents stated that the nearest medical facility to their home was primary Centre. 26.78% stated that private clinic was nearest medical health facility to their home. Only 14.28% stated that government hospital was nearest.

The study also revealed that 39.28% of the respondents choose treating doctor on basis of his consultation fee while 60.71% did not. As for the frequency of visits to a healthcare facility are concerned, 32.14% of the respondents visited the healthcare facility twice in a month, 28.57% of respondents twice in three months, 19.64% once in three months and 19.64% of respondents once in six months.

The 42.85% of the respondents stated that they have been ill in the past three months. The medicine taking behaviour of the respondents who were ill revealed that 58.92% took medicine on the physician's advice while 41.07% of the respondents depended on their past experience with the same illness.

Table 16.4 shows the first action taken by respondent when getting any health problem. The majority of the respondents (60.71%) will consult a pharmacist once they are experiencing any health problems. 30.35% of the respondents stated that they seek professional help. 3.57% of respondents practice self-medication as the first action if they face any health problem. 5.35% of the respondents will consult a traditional healer to treat their health problems.

Table 16.5 show the prevalence of self-medication in the respondents. 55.35% of the respondents reported that they do not practice self-medication while as 44.64% of the respondents reported that they do practice self-medication. In Shabri Mohalla 41.66% of respondents reported that they practice self-medication, in

Table 16.3 Utilization of medical system by respondents

Category	Percentage
Medical system preferred	90.16
Allopathy	
Homeopathy	6.27
Traditional healers	3.57
Medical institutions of first choice	
Near to home	21.42
Public	44.64
Private	55.34
Reason of choosing a particular health facility	
Good quality treatment	71.42
Affordable	7.14
Frequency of visits	
Twice in a month	32.14
Once in a month	28.57
Once in three months	19.64
Once in six months	19.64
Ill in past three months	
Yes	42.85
No	57.15
Nearest health facility to their home	
PHC/CHC	58.92
Government Hospital	14.28
Private clinic	26.78
Medicine taking Behaviour	
Physician advice	58.92
Self-medication	38.76
Home remedy	2.32
Seeks treatment	
Moderate symptoms	42.85
Mild symptoms	32.14
After few days of illness	25.00

Source: Based on Data obtained from Field work 2021.

Table 16.4 First action taken when getting any health problem

First action taken when you are not feeling well	Percentage
Consult a pharmacist	60.71
Consult a traditional healer	5.35
Consult a professional (physician)	30.35
Self-Medication	3.59

Source: Based on Data obtained from Field work 2021.

Table 16.5 Prevalence of self-medication

	Practice of self-medication (%age)	
Villages	Yes	No
Shabri Mohalla	41.66	58.34
Kankati	46.15	53.85
Sheikh Mohalla	50.00	50.00
Mir Mohalla	45.45	54.55
Kabutar Khana	40.00	60.00
Average	**44.65**	**55.35**

Source: Based on Data obtained from Field work 2021.

Kanketi 46.15%, in Sheikh Mohalla 50%, in Mir Mohalla 45.45% and in Kabutar Khana 40% of the respondents practice it.

Perception of Disease

Table 16.6 shows the perception of disease among the respondents. Only 41.07% of the respondents understood their health problem. Whether respondents knew the cause of their health problem, the study revealed that 53.57% of the respondents did not know the cause of their health problem while 46.42% of the respondents stated that they knew the reasons of their health problem.

Management of Diseases

Table 16.7 shows the management of health problems by the respondents. According to the results depicted in the table 73.21% of the respondents undertook general health screening while 26.78% of the respondents did not to screening. Similarly, only 44.64% of respondents undertook health-oriented activities to tackle their health problems.

Majority of respondents (98.21%) discussed their health problems with their family whereas only 1.78% of the respondents discussed with their friends.

The study also revealed that majority (66.07%) of the respondents followed the precautions prescribed by the doctor and 33.92% of the respondents did not follow them. The 42.85% of respondents consume the prescribed medicine until their symptoms are resolved. 39.28% of the respondents complete the course prescribed by the doctor. 17.85% of them leave it mid-way. The 67.85% of respondents stated that they gave up those habits or activities that caused or aggravated their health problems while as 32.14% did not gave up those habits.

Many studies suggest that addressing healthcare behaviour paves ways for appropriate utilization of healthcare services. The purpose of this study was to assess the healthcare behaviour that is utilization of medical systems, perception of diseases and management of health problems among the Hanji community of Dal Lake. The findings of this study show that utilisation of allopathic medical system

Table 16.6 Perception about diseases

Category	Percentage
Understand the health problem	
Yes	41.08
No	58.92
Understand the cause of health Problem	
Yes	46.43
No	53.57

Source: Based on Data obtained from Field work 2021.

Table 16.7 Management of health problems

Category	Percentage
Screened for general health	
Yes	73.21
No	26.79
Take health-oriented activities	
Yes	44.64
NO	55.36
Discuss health problems	
Family	98.21
Friends	1.79
Don't discuss	0.00
Follow precautions	
Yes	66.08
No	33.92
Treatment course completion	
Until Symptoms are resolved	42.85
Complete treatment Course	39.28
Leave it midway	17.85
Give up health causing aggravating/habits	
Yes	67.85
No	32.15

Source: Based on Data obtained from Field work 2021.

was high among the participants. The current study revealed that more participants preferred private clinics over the public as they provide better treatment. Also, the lack of medical staff and availability of medicines in public facility particularly in the primary centres could be a reason for participants to choose a private facility over the public. The study also revealed that majority (60.71%) would consult a pharmacist rather than a doctor as first action when getting any health problem. The reason could be that the pharmacists are easily available in the area. Another reason could be high consultation fee of doctors or the illness is not that severe.

Majority of participant sort treatment after a few days of illness rather on the onset of the illness. The reason could be that respondents believe that illness would either improve over time or improve on their own. Another reason could be that respondents think they are not sick enough to seek treatment.

Self-medication is defined as using medications without being supervised by healthcare providers. The revealed that about 44% of the participants practiced self-medication. The reason for self-medication among the participants could be to get quick relief from symptoms, self-diagnosis of the diseases or prior use of the medication. Only 41.07% of the participants understood their health problems. Similarly, only 46.42% understood the cause of their health issues. Lack of education could be the reason for the same. About 73.21% screened their general health which is satisfactory. Health oriented activities among the participants (44.64%) was low, it could be due to insufficient time to exercise, inconvenience of exercise, lack of self-motivation or non-availability of parks and sidewalks. Treatment course completion data of respondents showed that 42.85% participants consume until their systems are resolved and not complete the full course due to either high costs of medicine or feel that they have completely recovered.

Conclusion

The overall healthcare behaviour of the households was quite decent. According to their perception about their own health 41% of participants thought of their health as good. More than 50% preferred private clinics due to the good quality of treatment provided by them. Self-medication was also widely practised in the area.

Although majority of the participants waited few days after the onset of illness. Most of the participants took action when getting ill. Most of the participants tend to consult a pharmacist as first action to treat any health problems whereas some minority of participants would seek professional help as a first action if they faced any health issue. Participants utilized allopathic medical system more than any other system. In the study area it is essential to carryout awareness measures so as to reduce the practice of self-medication. The healthcare professionals and other healthcare authorities should work together to increase the awareness about the negative effects of self-medication in the area. Also, awareness should be conducted so that there is very less time gap between symptoms and seeking treatment. Also, there is urgent need to improve health facilities in the area. The government should try to strengthen the infrastructure of public health facilities. The government should also increase the manpower and availability of drugs in the public health facilities. The use of healthcare centres for proper diagnosis and consultation should be encouraged through proper information, education and communication.

Suggestions

- The literacy status of the people in the area needs to be improved and encouragement of quick health-seeking behaviour among the people. Proper health education and counselling should be provided by the health workers in the area.

- Healthcare infrastructure need to be improved in the area and Improve man power and availability of drugs in the nearest primary centres/dispensary/ hospitals.
- The diagnostic facilities should be made available in the area and tests for general screening should be provided free or at an affordable rate.

References

Abuduxike, G., Asut, O., Vaizoglu, S. A., & Cali, S. (2019) Health-seeking behaviours and its determinants: a facility-based cross-sectional study in the Turkish Republic of Northern Cyprus. *International Journal of Health Policy and Management, 9*(6), 240–249.

Adhikari, D., & Rijal, D. P. (2014). Factors affecting health seeking behaviour of senior citizens of Dharan. *Journal of Nobel Medical College, 3*(1), 50–57.

Ali, U. (2015). Impact of anthropogenic activates on Dal Lake (ecosystem/conservation strategies and problems). *International Journal of u-and e-Service, Science and Technology, 8*(5), 379–384.

Andersen, R., & Newman, J. F. (1973). Societal and individual determinants of medical care utilization in the United States. *The Milbank Memorial Fund Quarterly. Health and Society*, Winter; 51(1), 95–124.

Asfaw, L. S., Ayanto, S. Y., & Aweke, Y. H. (2018). Health-seeking behaviour and associated factors among community in Southern Ethiopia: Community based cross-sectional study guided by Health belief model. *BioRxiv*, 388769.

Begashaw, B., Tessema, F., & Gesesew, H. A. (2016). Health care seeking behaviour in Southwest Ethiopia. *PLoS One, 11*(9), e0161014.

Bhat, R. A., Shafiq-ur-Rehman, M. M., Dervash, M. A., Mushtaq, N., Bhat, J. I. A., & Dar, G. H. (2017). Current status of nutrient load in Dal Lake of Kashmir Himalaya. *Journal of Pharmacognosy and Phytochemistry, 6*(6), 165–169.

Bojola, F., Dessu, S., Dawit, Z., Alemseged, F., & Tessema, F. (2018). Assessment of Health Care Seeking Behaviour among House Hold Heads in Dale Woreda, Sidama Zone, Southern Ethiopia, Ethiopia. *Global Journal of Medical Research: FDiseases, 18*(1), 18–29.

Dawood, O. T., Hassali, M. A., Saleem, F., Ibrahim, I. R., Abdulameer, A. H., & Jasim, H. H. (2017). Assessment of health seeking behaviour and self-medication among general public in the state of Penang, Malaysia. *Pharmacy Practice (Granada), 15*(3). Sept. 30,p. 991.

Fazal, S., & Amin, A. (2012). Hanjis activities and its impact on Dal Lake and its environs (a case study of Srinagar City, India). *Research Journal of Environmental and Earth Sciences, 4*(5), 511–524.

Gupta, V. B. (2010). Impact of culture on healthcare seeking behaviour of Asian Indians. *Journal of Cultural Diversity, 17*(1), 13–19.

Inche Zainal Abidin, S., Sutan, R., & Shamsuddin, K. (2014). Prevalence and determinants of appropriate health seeking behaviour among known diabetics: results from a community-based survey. *Advances in Epidemiology*. November, *12*, DOI:10.1155/2014/793286.

Iyalomhe, G. B., & Iyalomhe, S. I. (2012). Health-seeking behaviour of rural dwellers in southern Nigeria: implications for healthcare professionals. *International Journal of Tropical Disease & Health, 2*(2), 62–71.

Kakai, H., Maskarinec, G., Shumay, D. M., Tatsumura, Y., & Tasaki, K. (2003). Ethnic differences in choices of health information by cancer patients using complementary and alternative medicine: an exploratory study with correspondence analysis. *Social Science & Medicine, 56*(4), 851–862.

Latunji, O. O., & Akinyemi, O. O. (2018). Factors influencing health-seeking behaviour among civil servants in Ibadan, Nigeria. *Annals of Ibadan Postgraduate Medicine, 16*(1), 52–60.

Macro, O. R. C. (2006). Central Statistical Agency: Ethiopia demographic and health survey 2005. *ORC Macro, Calverton, Maryland, USA.*

Oberoi, S., Chaudhary, N., Patnaik, S., & Singh, A. (2016). Understanding health seeking behaviour. *Journal of Family Medicine and Primary Care, 5*(2), 463.

Rahman, M., Islam, M. M., Islam, M. R., Sadhya, G., & Latif, M. A. (2011). Disease pattern and health seeking behaviour in rural Bangladesh. *Faridpur Medical College Journal, 6*(1), 32–37.

Ryff, C. D., & Singer, B. (1998). The contours of positive human health. *Psychological Inquiry, 9*(1), 1–28.

Vaishnavi, B., & Mishra, A. K. (2021). Health-seeking behaviour of patients with diabetes mellitus: a community-based cross-sectional study in an urban area of Pondicherry. *Journal of Current Research in Scientific Medicine, 7*(1), 33.

van der Hoeven, M., Kruger, A., & Greeff, M. (2012). Differences in health care seeking behaviour between rural and urban communities in South Africa. *International Journal for Equity in Health, 11*(1), 1–9.

World Health Organization. (1997). *Health Promotion Glossary of Terms.* Geneva. Accessed 30 December 2013.

17 Spatial Disease Pattern and its Correspondence with Geographical Factors of Jammu Region, J&K India

Subhash Chander Sharma

Introduction

The disease has assumed many shades ranging from sub-clinical to clinical Cases. Though disease is a departure from a state of health, yet it is very difficult to know where health ends and ill health begins. The disease could be diagnosed only when it produces significant changes in the body. From geographic angle, the disease is interpreted with respect to the coincidence in time and space of an agent and host (May 1950, Learmonth, 1988). An appropriate interaction of pathological factors (Pathogens, which causes disease) with respect to geographical factors (geogens environmental factors supporting pathogens) in relation to space and time, provide a better comprehension of the causes, which may be far deeper and more complex.

Objectives

Taking totality of health components the aims and objectives of the proposed study are outlined as under:

1 To investigate spatial variation of disease patterns.
2 To identify major "geogenic" and "pathogenic" determinants related to major diseases.
3 To examine the aspects of nutrition-deficiency diseases.
4 To highlight the disease intensity at tehsil level.
5 To delineate disease intensity zones for disease regionalization of the study area.

Sources of Data

The data have been obtained for the proposed study from primary as well as secondary sources. The secondary data for three years (1995–98) regarding health statistics consisting of incidence of diseases have been collected from the Directorate of Health, Office of Chief Medical Officer (CMO) and Block Medical Officer (BMO). General Statistics information have been taken from District Statistical Hand Books. The primary data pertaining to diseases, socio-economic, dietary

DOI: 10.4324/9781003329459-17

habits, living conditions, demographic components and health care behaviour and income level have been collected through field survey.

Methodology

Samples villages have been selected and are traversed across the longitudinal length of physiographic divisions. The secondary data are analysed at tehsil level. Absolute number of reported cases does not depict true level of disease incidence. For spatial analyses, secondary data have been analysed by using intensity rate (number of reported cases per 1,000 persons). The spatial distribution of 14 major diseases has been drawn on the basis of morbidity percentages and distribution levels of diseases have analysed by Rank-coefficient method. The pattern emerging over maps have been interpreted keeping in view the cause effect relationship.

Study Area

The study area lies between the Ravi river in the east and Actual Line of Control (ALC) in west Pakistan in the south, Kashmir Valley in the North and Ladakh lies to its north east. The area extends between 32° 17' and 34° 12' North latitudes and 73° 58' to 76° 47' East longitudes. Administratively Jammu Region is divided into six districts viz. Jammu, Kathua, Udhampur, Doda, Rajouri and Poonch. These districts further divided into 36 tehsils (Figure 17.1)

Disease Ecology

One of the most vital and significant approaches for analysing disease in the context of human environment interaction is that of disease ecology. Each disease and more often a group of disease has a specific pattern (Hazra, 1987).Here an attempt has been made to comprehend the spatial Pattern of diseases and its correspondence with geographical factors. The study lays more Emphasis on local adoption and effects, which form a basic parameter in understanding disease dynamic (Hughes and Hunter, 1970). The various agents which facilitate the spread of disease are examined in relation to human because he is the main contribution to the disease through his cultural traits. Therefore, man is a focal point and disease ecology acts as matrix of various environmental factors in associative occurrence.

Morbidity Index

It is the ratio between number of cases reported in an area and the number of expected cases. The expected cases are assumed to be in proportion to the ratio between total reported cases in a particular disease and that population of the region. Intensity of spatial distribution of disease in general context has been analysed by the following morbidity index formula (De Jayasree and Gollerkeri, 1984):

$$MI = \frac{OC}{EC} \times 100$$

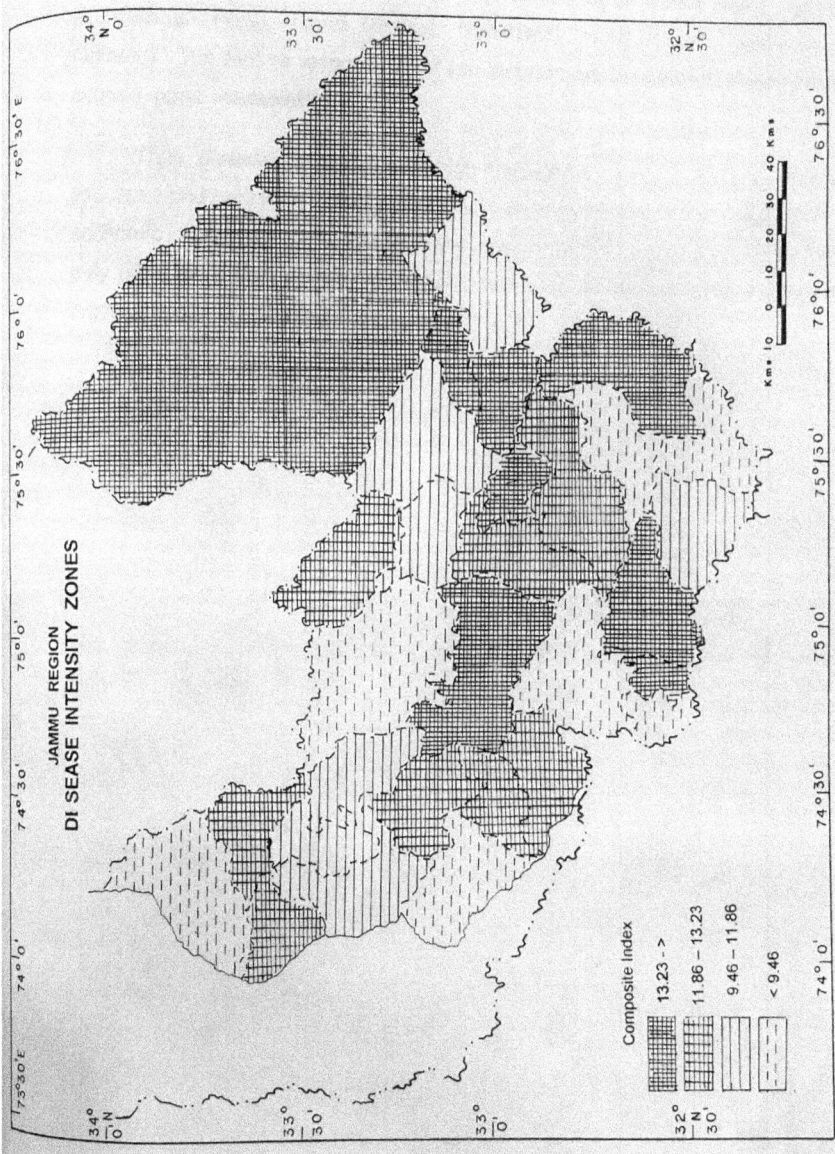

JAMMU REGION
DISEASE INTENSITY ZONES

Composite Index

13.23 - >

11.86 – 13.23

9.46 – 11.86

< 9.46

Figure 17.1 Administrative regions

Table 17.1 Morbidity index by tehsil

S.No.	Tehsil	Influenza	Whooping cough	Tuber-culosis	Measles	Mumps	Viral fever	Acute diarrhoea disease	Viral hepatitis	Typhoid fever	Helminthes disease	Inessential protozoal infection	Malaria	Eye disease	Teeth gum
1	Banihal	163.67	–	–	180.21	–	414.63	71.07	–	47.08	–	–	–	52.71	51.11
2	Ramban	105.42	–	7.30	139.22	14.259	45.81	50.80	18.84	619.30	117.61	284.76	–	63.71	71.08
3	Kishtwar	342.52	768.01	5.17	63.49	106.69	201.34	207.08	19.70	399.92	113.90	69.52	–	89.11	106.52
4	Gandoh	43.68	–	–	–	714.58	200.22	142.08	–	20.47	31.83	30.48	–	16.58	246.02
5	Bhaderwah	134.05	–	3.24	717.89	788.38	119.03	92.87	112.73	368.44	27.82	19.79	–	40.95	350.19
6	Thathri	134.05	–	3.24	717.89	788.38	119.03	92.87	112.73	368.44	27.82	19.79	–	40.95	350.19
7	Doda	120.86	–	54.28	369.75	129.58	58.81	57.14	25.21	102.21	15.43	55.06	6.22	74.97	135.59
8	Gool Gulab Garh	21.30	–	6.54	1.92	0.69	133.04	125.49	–	43.55	0.29	–	–	–	–
9	Reasi	142.11	–	93.56	38.58	110.04	94.98	78.68	327.20	29.16	91.16	35.81	–	166.96	248.52
10	Chenani	154.09	–	154.05	166.83	254.45	793.12	464.94	149.81	54.05	–	–	–	–	179.29
11	Ramnagar	165.55	18.23	165.50	95.51	128.99	71.69	49.56	8.69	147.37	158.13	136.58	119.17	156.90	74.22
12	Udhampur	106.00	–	105.97	12.57	110.07	61.27	46.82	879.21	89.80	81.07	145.91	–	142.37	51.74
13	Mendhar	44.60	348.44	44.59	40.30	106.79	20.57	130.01	44.20	117.46	176.77	184.91	–	–	51.74
14	Surankot	29.08	78.86	29.07	296.45	–	147.36	112.58	69.04	34.75	96.28	–	–	47.85	79.84
15	Poonch	73.46	–	73.43	55.25	96.07	34.46	109.32	37.45	16.21	18.13	23.12	38.03	168.80	76.43
16	Budhal	99.02	–	51.13	–	–	144.58	41.87	–	–	97.77	210.86	–	28.46	–
17	Nowshehra	6.85	–	6.85	9.87	35.48	34.30	31.72	112.24	76.42	–	–	–	–	152.28
18	Sunderbani	77.88	–	186.72	387.02	180.70	107.88	86.43	35.61	83.53	118.00	22.11	–	14.08	73.78
19	Kalakota	77.88	–	186.72	387.02	180.70	107.88	86.43	35.61	83.53	118.00	22.11	–	14.08	73.78
20	Thanamandi	14.34	84.42	159.69	124.60	66.24	21.50	181.51	171.57	227.81	127.76	62.13	172.21	18.79	33.07

(Continued)

Table 17.1 (Continued)

S. No.	Tehsil	Influenza	Whooping cough	Tuber-culosis	Measles	Mumps	Viral fever	Acute diarrhea disease	Viral hepatitis	Typhoid fever	Helminthes disease	Inessential protozoal infection	Malaria	Eye disease	Teeth gum
21	Rajauri	14.34	84.42	159.69	124.60	66.24	21.50	181.51	171.57	227.81	127.76	62.13	172.21	18.79	33.07
22	Akhnoor	55.90	227.54	55.89	144.63	160.49	75.12	106.26	12.27	24.55	449.20	3.01	–	69.24	124.42
23	Ranbirsingh Pora	38.50	–	102.99	9.22	–	–	86.49	22.04	27.51	67.61	52.52	–	113.09	86.63
24	Bishna	90.87	–	100.52	-	–	290.87	143.80	–	5.73	83.88	118.90	–	522.55	130.30
25	Samba	122.10	–	323.91	13.86	84.67	51.18	144.98	77.37	34.31	118.90	277.01	–	266.04	45.98
26	Jammu	29.75	–	88.71	-	49.79	67.67	48.06	2.69	11.47	87.18	224.97	145.92	36.69	0.06
27	Billawar	12.63	–	66.75	21.38	72.55	106.63	55.31	70.25	147.09	39.25	54.45	–	110.49	83.31
28	Bashohli	306.13	–	19.31	216.88	95.01	84.00	71.57	57.62	253.53	251.85	205.14	–	224.72	86.31
29	Hiranagar	97.00	149.52	99.62	158.52	12.96	222.71	112.63	29.39	5.23	52.18	16.43	–	224.72	86.31
30	Kathua	30.03	645.28	119.01	4.80	3.45	143.54	66.68	3.54	0.31	18.25	6.05	72.13	124.31	261.49

Source: Field Work by Author.

where MI = Morbidity Index, OC = Observed Cases, EC = Expected case

Table 17.1 shows the distribution of disease by morbidity at tehsil level. On the basis of mean(X) and Standard deviation (σ), four levels of disease morbidity have been classified, like high morbidity, moderately high morbidity, moderately low morbidity and low morbidity. For selected disease for want of adequate and reliable data, the relationship between diseases and environmental factors, in depth has led to inferences on the basis of which the related map has been prepared.

Classification of Diseases

Most of the diseases in Jammu region are of infectious nature, associated with sub-tropical to sub-temperate climatic conditions. Mode of transmission of diseases under different geographical environment has been taken the basis of classification as they are related to pathogens involved in spreading the diseases. The main diseases of study area have been classified into two broader groups as under:

A (I Communicable Diseases (Secondary Sources)
(i) Acute Respiratory Infections
Influenza (b) Whooping Cough (c) Tuberculosis (d) Measles (e) Mumps
(f) Viral fever
(ii) Intestinal Infection
(a) Acute Diarrhoeal Diseases (b) Viral Hepatitis (c) Typhoid Fever (d) Helminthic Diseases (worm Infection) (e) Intestinal Protozoal Infection (Amoebiasis and Giardiasis)
(iii) Arthropod borne Diseases
Malaria
A (II) Deficiency Diseases
Eye Diseases (b) Teeth and Gum ailments
B Non Communicable Diseases (Primary Sources)

Communicable Disease

The communicable diseases are those in which disease causative agents are transmitted directly or indirectly from patients to a susceptible person, animal to human, animal to animal through environmental factors. These communicable diseases include contagious as well as infection diseases. There are six basic groups of organisms viz. viruses, bacteria, protozoa, insects, helminths and fungi; and five main modes of transmission like physical contact air, food and water, insects and commensals. The communicable diseases of study area have been discussed below under different sub-headings.

Acute Respiratory Infection

A group of acute diseases due to inflammation of the respiratory tract anywhere from nose to alveoli cause acute respiratory infections (ARIs). The major ARIs diseases recorded by health department in Jammu region are discussed as under.

Influenza

It is an acute respiratory tract infection caused by influenza viruses, namely influenza type A, B and C or arthomyxovirida family. Influenza is the most dominated disease in the region. It has 78.74 cases/1000 persons intensity rate. The distribution pattern reveals that disease has close correspondence to physical factors, Figure 17.2a. Tehsils Kishtwar and Basholi display high morbidity (176.79>) while Rajouri, Thanamandi, Billawar, Nowshehra have very low morbidity (<15.87). The moderately high morbidity (97.33–176.79) has been recorded in Ramnagar, Banihal, Chenani, Reasi, Bhaderwah, Thathri, Samba, Doda, Udhampur, Ramban and Budhal tehsils. The tehsils of Hiranagar, Bishna, Sunderbani, Kalakote, Punch, Akhnoor, Mendhar, Gandoh, Ranbirsingh Pora, Kathua, Jammu, Surankote and GoolGulabGarh have recorded moderately low morbidity (15.87–97.33), (Figure 17.2a). Relative relief, rainfall, forest cover, poor health education and poverty and the main responsible environmental factors.

Whooping Cough

It is an air borne acute respiratory disease, caused by *Bacillus pertussis* or *Hemophiluspertiussis.* The disease has high morbidity rate among the age group below five years but more prone to female children. In spite of the fact that disease is widely prevalent in the region but only health centres of eight tehsils have reported whooping cough cases. Among these tehsils, Kishtwar and Kathua have recorded high morbidity (541.15>), while Mendhar has moderately high morbidity (290.3–541.15). The areas of moderately low morbidity (38.91–290.3) are Akhnoor, Hiranagar and Rajouri while low morbidity is (<38.91) found in Surankote and Ramnagar (Figure 17.2b). Long winters, unhygienic conditions, poverty, low female literacy, lack of health care consciousness, and malnutrition are responsible factors for high morbidity.

Tuberculosis

This infectious disease is caused by *Mycobacterium* tuberculosis. Tuberculosis is fairly prevalent in the entire region, with an intensity rate of 3.78 cases/1,000 persons, lower than national four cases/1,000 persons (Park, 2000). High to moderately high morbidity (84.25>) is confined in Samba, Sunderbani, Kalakote, Ramnagar, Rajouri, Thanamandi, Chenani, Kathua, Udhampur, RanbirsinghPora, Bishna, Hiranagar, Reasi and Jammu tehsils (Figure 17.2c). These tehsils are located in Foot Hill Plains and Siwalik hills. Tehsils of Punch, Billawar, Akhnoor, Doda, Budhal, Mendhar, Surankote, Bashohli form areas of moderately low morbidity (11.99–84.25). Tehsils of Ramban, Kishtwar, Nowshehra, GoolGulabGarh, Bhaderwah and Thathri exhibit low morbidity (<11.99). The disease is primarily attributed to cultural and socio-economic factors like room density, chronic alcoholism, smoking, poverty and lack of health awareness.

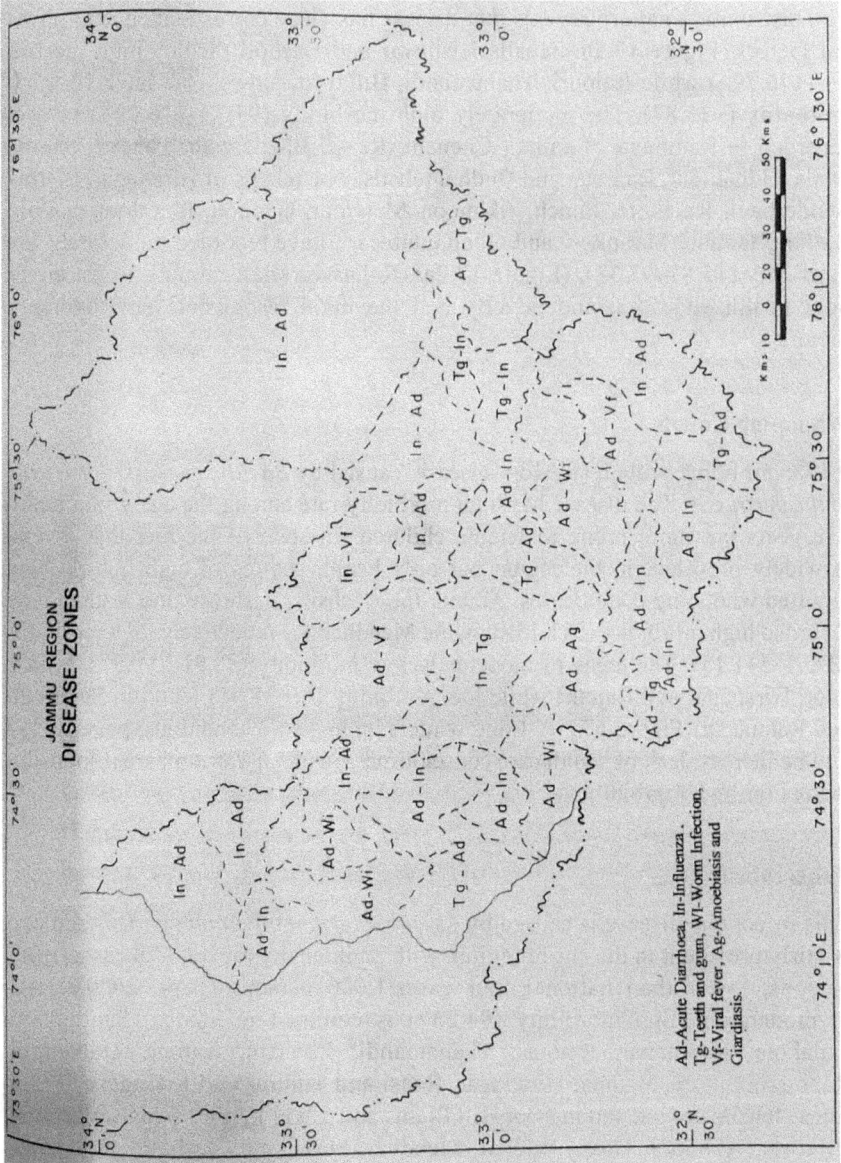

JAMMU REGION
DI SEASE ZONES

Ad-Acute Diarrhoea, In-Influenza.
Tg-Teeth and gum, Wl-Worm Infection,
Vf-Viral fever, Ag-Amoebiasis and
Giardiasis.

Figure 17.2 Pattern of respiratory borne diseases Morbidity index 1988–98

Measles

It is highly infectious disease caused by *Morbillivirus*. The data analysis shows its intensity rate of 0.38 cases /1,000 persons. High morbidity pattern (308.13>) is found in Bhaderwah, Thathri, Sunderbani, Kalakote and Dodathesils. The tehsils of Surankote, Bashohli, Banihal, Chenani, Hiranagar, Akhnoor have recorded moderately high morbidity (142.12–308.13). Low to moderately low morbidity (<59.67) has been registered in Punch, Mendhar, Reasi, Billawar, Samba, Udhampur, Nowshehra, RanbirsinghPora, Kathua, GoolGulabGarh, Ramban, Rajouri, Thandamandi, Ramnagar and Kishtwar (Figure 17.2d). Most of the mountainous areas depict high morbidity. Low temperature, poor sanitation, malnourishment and under nourishment of children are the main causes.

Mumps

It is an acute infectious disease caused by *Myxovirusparotidits*. Though it affects largely children in the age group 5–15 years but the disease is more severe among adults. The regional average of mumps disease intensity rate is 1.05 cases/1,000 persons. Figure depicts that Bhaderwah, Thathri and Gandoh tehsils from an area of high morbidity (346.71>). Moderately high morbidity (146.45–346.71) has been recorded in Chenani, Sunderbani, Kalakote, and Akhnoor tehsils. The area characterized with moderately low morbidity (46.81–146.45) consists of Doda, Ramnagar, Udhampur, Reasi, Mendhar, Kishtwar, Punch, Bashohli, Samba, Billawar, Rajouri, Thannamandi, while Jammu, Nowshehra, Ramban, Hiranagar, Kathua and GoolGulabGarh show low morbidity (<46.81). On the other hand tehsils of Banihal, Surankote, Budhal, RanbirsinghPora and Bishna have not reported any mumps incidence during 1995–98. Figure 17.2e reflects that except tehsil Akhnoor all other tehsils project high to moderate high morbidity (146.45>).

Viral Fever

Viral fever refers to heterogenous group of diseases which are related to spread of virus which attacks on specific organs of body. The general symptoms associated with the diseases are headache, bodyache, nasal congestion, cough, skin rash and muscle as well as joint pains. Viral fever affects all age groups of both sexes and its incubation period varies from days to several weeks. Most of the viral infections are of mild intensity but these sometimes turn to be fatal.

The viral fever is a fourth rank disease in Jammu region, showing 30.21 cases/1,000 persons intensity rate. Moderately high to high morbidity (144.10>) indicates in the area of Chenani, Banihal, Bishna, Hiranagar, Kishtwar and Gandoh tehsils. The pattern of moderately low morbidity (65.62–144.10) is observed in Surankote, Budhal, Kathua, GoolGulabGarh, Bhaderwah, Thatri, Sunderbani, Kalakote, Billawar, Reasi and Bashohli tehsils. Area of Akhnoor, Ramnagar, Jammu Udhampur, Doda, Samba, Ramban, Punch, Nowshehra, Rajouri, Thannamandi and Mendhar tehsils exhibit low morbidity (<65.62), (Figure 17.2f). The various physical environmental factors responsible for this disease.

Intestinal Infection

The nourishment of all parts of body is associated with our digestive system. It consists of small and large intestines. Water is the most dominating means of transmission of intestinal infection because it is used by every one in relatively more quantity than any other food products. The major intestinal infection diseases reported in Jammu region are as follows.

Acute Diarrhoeal Disease

It remains one of the most public health problems in the world. An increase in the frequency and fluidity of bowl movement relative to the usual pattern of each individual is defined as diarrhoea. Acute diarrhoea includes both diarrhoea and dysentery.

The incidence of diarrhoea disease is closely associated with warm and humid climate. Other responsive environmental factors which propagate the disease are under nutrition, poor sanitation, unhygienic living conditions, unsatisfactory methods of food preservation illiteracy, poor personal hygiene, etc. Therefore, acute diarrhoeal is generally known as the disease of unwashed.

The disease is widely and frequently reported in Jammu region with the recorded intensity rate of 111.51 cases/1,000 persons. Only two tehsils of Chennai and Kishtwar have registered high morbidity (190.45>). Moderately high morbidity (108.06–190.45) pattern has been found in Rajouri, Thanamandi, Samba, Bishna, Gandoh, Mendhar, GoolGulabGarh, Hiranagar, Surankote, and Punch tehsils (Figure 17.3a). Remaining tehsils show moderately low morbidity (<108.06). The study reveals that water scarcity as well as water surplus areas have the same pattern. It is attributed to the majority of people taking drinking water from natural resources, i.e. 63.32% in rural areas.

Viral Hepatitis

It is a highly communicable disease caused by Hepatitis viruses. WHO (1975) identified two main hepatitis viruses, like *Hepatitis* A and *Hepatitis* B.

The viral hepatitis A infection has been found to predominate in the region. The disease on an average has incidence rate of 1.53 cases/1,000 persons. Figure 17.3b, presents 14 tehsils of moderately low morbidity (11.80–103.94) and only two tehsils viz. Udhampur and Reasi of high morbidity (196.08 - >). The tehsils of Rajouri, Thanamandi, Chenani, Bhaderwah, Thathri and Nowshera form areas of moderately high morbidity (103.94–196.08) while low morbidity (<11.80) is found in Ramnagar, Kathua and Jammu tehsils.

Typhoid Fever

It is an acute infectious water borne disease caused by *Salmonella typhi* found only in human. Though disease occurs throughout the year but maximum typhoid incidence is reported during July – September, the period which coincides with rainy season. Typhoid infection has 10–14 days incubation period.

The study area registers 3.91 cases/1,000 persons intensity rate. Figure 17.3c reveals its high to moderately high morbidity (114.90>) pattern concentration in Ramban, Kishtwar, Bhaderwah, Thathri, Bashohli, Rajouri, Thanamandi,

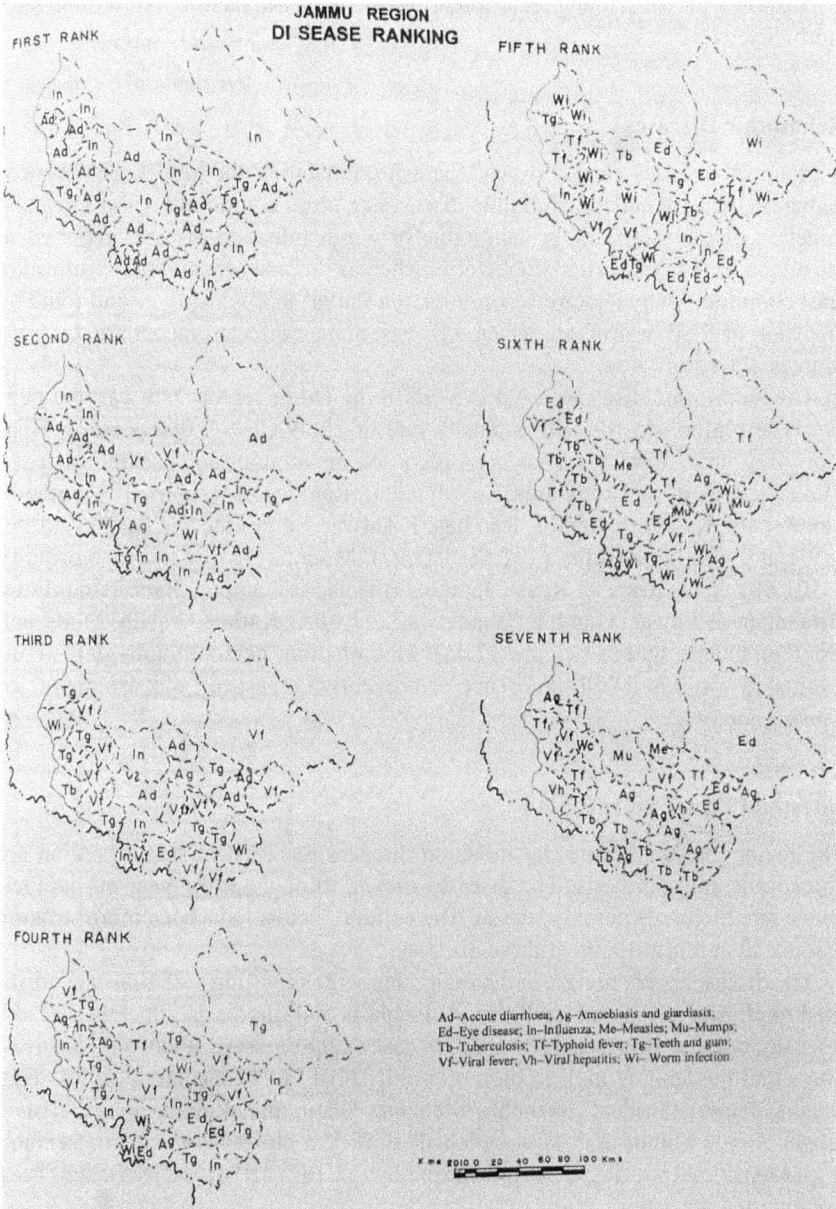

Figure 17.3 Pattern of intestinal diseases morbidity index 1988–98

Ramnagar, Billawar and Mendhar tehsils. The tehsils of Surankote, Samba, Reasi, RanbirsinghPora, Akhnoor, Gandoh, Punch, Jammu, Bishna, Hiranagar and Kathua are characterized with low morbidity (<41.98). The factors favourable for the growth of flies seem to be the main environmental agent in growth of disease. The poor sanitation, unsafe drinking water, unhygienic food habits, improper night soil disposal, personal unhygienic conditions are the basic factors contributing high morbidity.

Helminthic Diseases

A group of diseases caused by endoparasite, inhabiting the alimentary canal of human beings are called helminthic diseases or more commonly warm intestinal infection diseases. The main helminthic or worm infection diseases reported in the region are ascariasis and hookworm, whereas threadworm and tape worms are least common. Many acquire worm infection (larva) through skin or oral route by ingestion of food or drinking water, whereas, human infected faeces are the main sources of transmission.

The helminthic diseases are prevalent in the entire region with varying density. The region records with intensity rate of 25.09 cases/1,000 persons. High morbidity index (197.97>) has been recorded in Akhnoor and Bashohli tehsils. Eleven tehsils of the study area, i.e. Mendhar, Ramnagar, Rajouri, Thanamandi, Sunderbani, Kalakote, Samba, Ramban, Kishtwar, Surankote and Budhal exhibit moderately high morbidity (101.68–197.97). Moderately low to low morbidity (<101.68) concentrate in Reasi, Jammu, Bishna, Udhampur, RanbirsinghPora, Hiranagar, Billawar, Gandoh, Bhaderwah, Thathri, Kathua, Punch, Doda and GoolGulabGarh tehsils (Figure 17.4a). Malnutrition, personal unhygiene, contaminated water and soil, ignorance of preventive measures, etc. are the major causing factors.

Intestinal Protozoal Infection

The major intestinal protozoal infection diseases recorded in Jammu region are amoebiasis and giardiasis. The diseases prevail throughout the year but its incidence rate increases in rainy season. The cultural factors have been more strongly associated in transmission of these diseases.

The diseases predominate in the region but wide reporting variations are attributed to clinical diagnostic facilities. Amoebiasis and giardiasis infection diseases intensity rate has been recorded at 12.28 cases/1,000 persons. Figure 17.4b, illustrates that moderately high to high morbidity (101.76>) concentrate in Ramban, Samba, Jammu, Budhal, Bashohli, Mendhar, Udhampur, Ramnagar and Bishna tehsils. While Kathua and Akhnoor tehsils show low morbidity (<12.16). Remaining 14 tehsils show moderately low morbidity (12.16–101.76) and five tehsils have not reported any case.

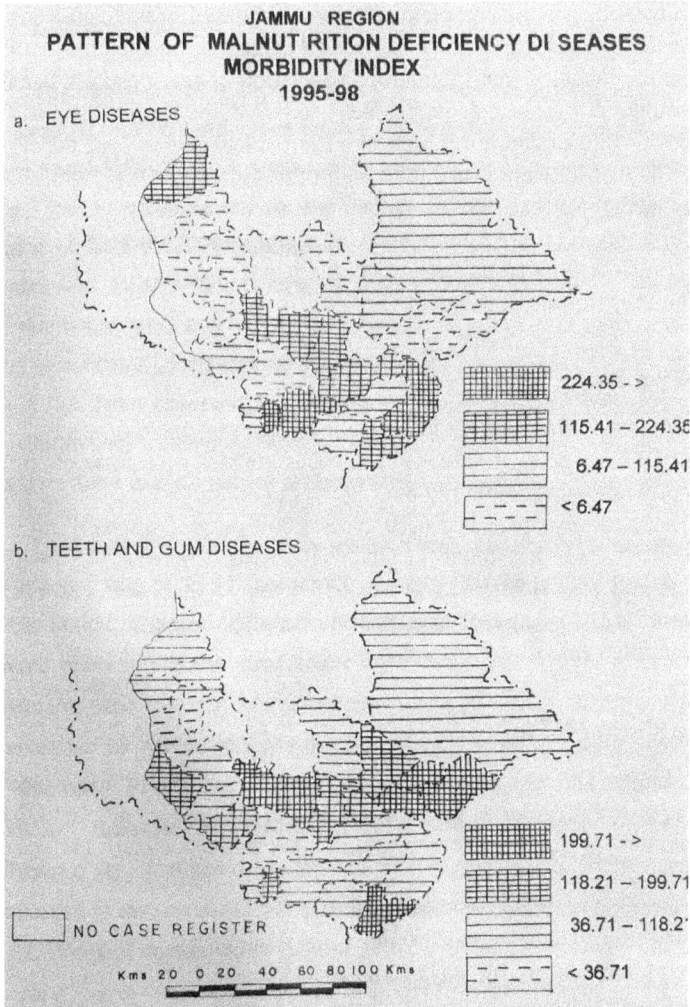

JAMMU REGION
PATTERN OF MALNUT RITION DEFICIENCY DI SEASES
MORBIDITY INDEX
1995-98

a. EYE DISEASES

224.35 - >
115.41 – 224.35
6.47 – 115.41
< 6.47

b. TEETH AND GUM DISEASES

199.71 - >
118.21 – 199.71
36.71 – 118.2
< 36.71

NO CASE REGISTER

Kms 20 0 20 40 60 80 100 Kms

Figure 17.4 Pattern of diseases of intestinal and protozoal infection 1988–98

Arthropodal Disease (Malaria)

The arthropoda phylum acts as carrier agents of infectious parasite from source to susceptible person. Malaria is the main arthropodal transmitted diseases in Jammu region.

The prevailing of malaria disease is associated with both physical and cultural environmental factors. The physical elements, i.e. temperature, rainfall, altitude have an overriding influence on the occurrence of the disease. Temperature between 20° and 30°C, 60% relative humidity and altitude not above 2,000–2,500 metres above sea level are complementary to mosquitoes growth and consequently malaria disease. The cultural factors such as irrigation channel, agriculture practices, poor drainage, water stagnation, vegetation cover, people's behaviour in sleeping outdoor without mosquitoes net, etc. affect the disease pattern. The disease intensity rate of Jammu region has been calculated to be 2.80 cases/1,000 persons. The district Rajouri exhibits high morbidity (151.08>), while Doda low morbidity with less than 33.48. Moderately high morbidity (92.28–151.08) of malaria reporting areas are Jammu and Udhampur district whereas it is moderately low (33.48–92.28) in Kathua and Punch districts (Figure 17.4c). The level of irrigation, rainfall, altitude, forest cover are positively correlated.

Field Survey Data

It has been observed that in addition to above communicable diseases, skin infection and Ear, Nose and Throat (ENT) diseases are also prominent communicable ailments prevailing in Jammu region.

Skin diseases have recorded 6.20 cases/1,000 persons intensity rate. Highest intensity 11.65 cases has found in Outer Himalaya, whereas 7.47 cases in Foot Hill Plains and only 0.80 cases in Lesser Himalaya have been recorded. These skin diseases are caused by viruses, bacteria, fungi, parasites and plants. High temperature supplemented by moisture is a significant climatic element in spreading skin infection like fungus infection (ringworm), parasitic infestations (scabies), skin allergy, etc. *Parthenium hysterophorus* commonly known as congress grass of family Asteraceae plant is causing severe skin allergies in Outer Himalaya and Foot Hill Plains. Other skin diseases found in study area are pimples, prickly heat, rashes, eczema, etc. As far as venereal diseases are concerned proper data have not been maintained by the health centre and primary survey also fails to trace any such case, for obvious reasons.

Jammu region records 2.96 cases/1,000 persons of ENT diseases. Outer Himalaya has registered highest intensity rate (4.66 cases), followed by Foot Hill Plains (3.74 cases) and Lesser Himalaya (0.80 cases). Ear infection, deafness, ear ache, sinusitis, throat soaring, tonsils and adenoids are main ENT diseases found in the study area. Common cold, influenza, high temperature, seasonal changes, unhygienic surroundings, dust allergy are the main factors responsible for such diseases.

Deficiency Diseases

Eye Diseases

The major eye diseases recognized in Jammu region are cataract, glaucoma, trachoma and conjunctivitis. The eye diseases are universally distributed, and its geographical variations are attributed to clinical factors. Jammu region has registered 13.36 cases/1,000 persons intensity rate. The tehsils Bishna, Samba, Bashohli, Punch, Reasi, Kathua, Ramnagar and Udhampur form an area of high to moderately high morbidity (115.41>). Seven tehsils viz. RanbirsinghPora, Billawar, Hiranagar, Kishtwar, Doda, Akhnoor and Ramban record moderately low morbidity (6.47–115.41) while Banihal, Surankote, Bhaderwah, Thathri, Jammu, Budhal, Rajouri, Thanamandi, Gandoh, Sunderbani and Kalakote tehsils exhibit low morbidity (<6.47) respectively (Figure 17.5a). Poor socio-economic status, lack of primary health care services, treatment by quacks, putting kajal and eye stick, personal unhygiene, vitamin A deficiency, etc. are factors responsible for various eye disease.

Teeth and Gum Diseases

The diseases include dental caries, pyorrhea and periodontal. They are chronic in nature and recognized as non-communicable diseases. These diseases take place due to deficiency of multiple nutrients like calcium, Vitamin A, ascorbic acid (vitamin C) and vitamin D but the deficiency of calcium and vitamin C plays a more crucial role in the widespread teeth and gum diseases (Farooqi, 1977; Agnihotri, 1995).

The teeth and gum diseases are third rank diseases in study area, which form intensity rate of 36.91 cases/1,000 persons. Tehsils of Bhaderwah, Thathri, Kathua, Reasi, Gandoh, Chenani, Nowshehra, Udhampur, DodaBishna and Akhnoor present high to moderately high morbidity (118.21>). Rajouri Thanamandi and Jammu tehsils record low morbidity (<36.71). The areas characterized by moderately low morbidity (36.71–118.21) distribution pattern are Kishtwar, Ranbirsingh Pora, Bashohli, Hiranagar, Surankote, Punch, Sunderbani, Kalakote, Ramnagar, Ramban, Billawar, Mendhar Banihal, Samba tehsil (Figure 17.5b). The low morbidity reported in Jammu and Rajouri tehsils may be attributed to maximum reporting to private clinics. Variable responsible for the high incidence of diseases are ground water, soil, non-vegetarian dietary habits, poor oral hygiene, less consumption of necessary vitamins, etc.

Field Survey Data

The primary data reveal that anaemia, paralysis and gout are other major nutrition related diseases found in the study area. Region records 8.89 cases /1,000 persons of anaemia which results due to deficiency of iron, folic acid and vitamin B_{12}. Highest (18.65 cases/1,000 persons) intensity rate has been found in Outer Himalaya followed by 6.85 cases in Foot Hill Plains and 4.81 cases in Lesser Himalaya (Table 17.2). Low agricultural production, and/or low economic level, imbalanced dietary habits, low consumption of vegetables are main causes for spreading anaemia.

Table 17.2 Disease intensity rate of communicable and deficiency diseases

Disease	Lesser Himalaya	Outer Himalaya	Foot Hill plains	Jammu region
Ear, nose, throat	0.80	4.66	3.70	2.96
Skin disease	0.80	11.65	7.47	6.20
Anaemia	4.81	18.65	6.85	8.89
Paralaysis	1.60	2.33	3.74	2.69
Gout	2.41	1.16	2.49	2.16

Source: Field Survey.

The paralysis and gout diseases show the same trend. Their intensity rates have been recorded as 2.69 and 2.16 cases/1,000 persons respectively. Factors responsible for paralysis are hypertension, diabetes, smoking, etc. Gout is the disease caused by defective production or action of insulin, a hormone that controls glucose, fat and amino acid metabolism. The body functioning and dietary habits are the main causes for gout.

Non-Communicable Diseases

A group of disease closely related to changing life style, naturally contaminated water with trace elements, urbanization, rapid industrialization, old age, etc., mostly affect middle- or old-age population. The intensity rate of such diseases has been found to be high in males than females. Epidemiologists have recognized six sets of 'risk factors' responsible for major non-communicable disease (Park, 2000), such as smoking, alcohol abuse, wrong dietary pattern, environmental risk factors (e.g. occupational hazards etc.), non-acceptance of preventive health measures (e.g. hypertension, diabetes control) and stress factors. Like infectious diseases, chronic diseases pathogens also require a coincidence in time and space of agent and host. The major difference, however, is that that diseases analysis is more complex because non-single factor associative to diseases. Non-Communicable diseases analysis is more complex because no single factor associative to disease may be discussed in isolation.

Field data show that region has low non-communicable disease intensity as compared to national level. It is attributed that most of the surveyed population belongs to rural areas. The intensity rate of non-communicable diseases of study area has been calculated and shown in the Table 17.3. Arthritis, a first rank disease has been recorded in the region and has 22.37 cases/1,000 persons, showing maximum reporting (34.87 cases) in Foot Hill Flains. Second rank disease, the chronic asthma has 15.36 cases/1,000 persons and shows high concentration in Foot Hill

Table 17.3 Disease intensity rate of non-communicable diseases

Disease	Lesser Himalaya	Outer Himalaya	Foot Hill plains	Jammu region
Arthritis	11.23	15.15	34.87	22.37
Asthma	12.04	13.99	18.68	15.36
Hypertension	12.04	15.15	13.70	13.48
Cardiovascular	4.81	2.33	6.85	5.12
Diabetes	1.60	2.33	2.49	2.12
Cancer	–	–	1.88	0.81
Stone (kidney & gall bladder)	8.83	3.50	6.85	6.74
Headache	–	1.16	8.72	4.04
Mental disorder	–	2.33	9.34	4.58
Piles	–	1.16	1.88	1.08
Others (tension, allergy, fits)	–	–	3.11	1.35

Source: Field Survey.

Plains. Hypertension register the third rank, and shows 13.48 cases/1,000 persons. Outer Himalaya shows slightly high intensity rate (15.15 cases/1,000 persons) than Foot Hills Plains (13.70 cases/1,000 persons). Other non-communicable diseases like cardiovascular diseases, diabetes, cancer, stone aliments, headache, mental diseases, piles, allergic diseases have also been recorded in the region, which show below ten cases/1,000 persons intensity rate (Table 17.3).

It is surprising to note that Lesser Himalaya records significant intensity of cardiovascular and stone ailment diseases in the region. It has been observed that Foot Hill Plains has recorded high intensity in non-communicable diseases. The changing way of life, improper dietary habits, fertilizer contaminated trace elements in drinking water, smoking, consumption of local alcohol in villages, indoor air pollution, dust and pollen allergies, hereditary tension or stress, old age, etc. are the main environmental factors contributing non-communicable diseases in the region.

Ranking of Diseases

The geographical distribution pattern of diseases has been analysed by disease ranking method. It depicts the relative dominance of different diseases in general morbidity pattern of a region. The disease ranks have been determined by taking percentage share of each disease and results represented cartographically (Figure 17.5).

Table 17.4, reveals that acute diarrhoea rank first in Chenani, Mendhar, Sunderbani, Kalakote, Bishna, Samba, Hiranagar, Ramnagar, Rajouri, Thanamandi, Akhnoor, Jammu, GoolGulabGarh, Billawar, Gandoh and RanbirsinghPora, influenza in Banihal, Ramban, Kishtwar, Doda, Punch, Surankote, Budhal, Reasi, Bashohli, and teeth and gum in Bhaderwah, Thathri, Nowshehra, Udhampur and Kathua tehsils.

Second ranking disease shows much-diversified pattern. Tehsils Ramban, Kishtwar, Doda, Udhampur, Surankote, Punch, Budhal, NowshehraBashohli and Kathua have diarrhoea as a second rank disease. While influenza in Bhaderwah, Thathri, Chenani, Mendhar, Sunderbani, Kalakote, Bishnah, Samba and Hiranagar. Worm infection as second-rank disease is found in Ramnagar, Rajouri, Thannamandi, Akhnoor; teeth and gum in Gandoh, Reasi and RanbirsinghPora; viral fever in Banihal, GoolGulabGarh and Billawar; amoebiasis and giardiasis in Jammu tehsils.

Third rank illustrates further more diversification. Teeth and Gum stands third rank disease in Doda, Ramnagar, Punch, Rajouri, Thannamandi, Akhnoor and Billawartehils while viral fever in Kishtwar, Gandoh, Chenani, Udhampur, Surankote, Budhal, Sunderbani, Kalakote, Bishna, Hiranagar and Kathua, Tehsils Banihal, Bhaderwah, Thathri and Reasi have diarrhoea as third rank disease; influenza in Gool Gulab Garh, Ranbirsingh Pora in Jammu; worm infection in Mendhar and Bashohli; amoebiasis and giardiasis in Ramban; eye diseases in Samba tehsils. Similarly, the rank distribution of diseases of IV, V, VI and VII exhibits mixed pattern. The study depicts that three diseases dominate in the region. Acute diarrhoea concentrates in first three ranks; influenza in first five; teeth and gum in first six. Further, study reveals that none of the disease hold all the seven ranks. Another disease which hold maximum six ranks (from II to VII) is viral fever (Table 17.4).

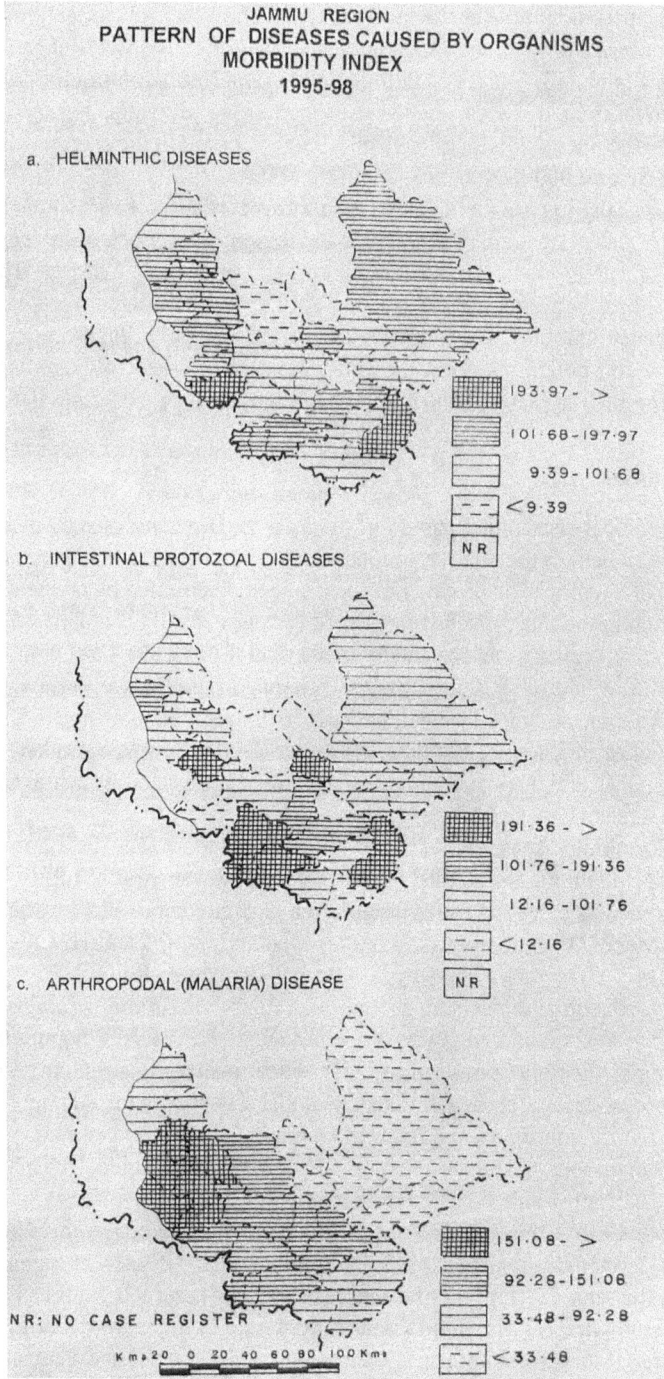

JAMMU REGION
PATTERN OF DISEASES CAUSED BY ORGANISMS
MORBIDITY INDEX
1995-98

a. HELMINTHIC DISEASES

193·97 - >
101·68 - 197·97
9·39 - 101·68
< 9·39
N R

b. INTESTINAL PROTOZOAL DISEASES

191·36 - >
101·76 - 191·36
12·16 - 101·76
< 12·16
N R

c. ARTHROPODAL (MALARIA) DISEASE

151·08 - >
92·28 - 151·08
33·48 - 92·28
< 33·48

NR: NO CASE REGISTER

Km 20 0 20 40 60 80 100 Kms

Figure 17.5 Pattern of malnutrition/deficiency diseases morbidity index 1988–98

Table 17.4 Ranking of diseases

S.No.	Tehsils	I	II	III	IV	V	VI	VII
1	Banihal	In	Vf	Ad	Tg	Ed	Tf	Me
2	Ramban	In	Ad	Ag	Wi	Tg	Tf	Vf
3	Kishtwar	In	Ad	Vf	Tg	Wi	Tf	Ed
4	Gandoh	Ad	Tg	Vf	In	Wi	Mu	Ag
5	Bhaderwah	Tg	In	Ad	Vf	Tf	Wi	Ed
6	Thathri	Tg	In	Ad	Vf	Tf	Wi	Ed
7	Doda	In	Ad	Tg	Vf	Ed	Ag	Tf
8	GoolGulabGarh	Ad	Vf	In	Tf	Tb	Me	Mu
9	Reasi	In	Tg	Ad	Vf	Wi	Ed	Ag
10	Chenani	Ad	In	Vf	Tg	Tb	Mu	Vh
11	Ramnagar	Ad	Wi	Tg	Ed	In	Vf	Ag
12	Udhampur	Tg	Ad	Vf	In	Wi	Ed	Ag
13	Mendhar	Ad	In	Wi	Ag	Tg	Vf	Tf
14	Surankote	In	Ad	Vf	Tg	Wi	Ed	Tf
15	Punch	In	Ad	Tg	Vf	Wi	Ed	Ag
16	Budhal	In	Ad	Vf	Ag	Wi	Tb	Wc
17	Nowshehra	Tg	Ad	Tb	Vf	In	Tf	Vh
18	Sunderbani	Ad	In	Vf	Tg	Wi	Tb	Tf
19	Kalakote	Ad	In	Vf	Tg	Wi	Tb	Tf
20	Thanamandi	Ad	Wi	Tg	In	Tf	Tb	Vf
21	Rajouri	Ad	Wi	Tg	In	Tf	Tb	Vf
22	Akhnoor	Ad	Wi	Tg	In	Vf	Ed	Tb
23	Ranbirsingh Pora	Ad	Tg	In	Wi	Ed	Ag	Tb
24	Bishna	Ad	In	Vf	Ed	Tg	Wi	Ag
25	Samba	Ad	In	Ed	Ag	Wi	Tg	Tb
26	Jammu	Ad	Ag	In	Wi	Vf	Tg	Tb
27	Billawar	Ad	Vf	Tg	Ed	In	Wi	Ag
28	Bashohli	In	Ad	Wi	Tg	Ed	Ag	Vf
29	Hiranagar	Ad	In	Vf	Tg	Ed	Wi	Tb
30	Kathua	Tg	Ad	Vf	In	Ed	Wi	Tb
	Jammu region	Ad	In	Tg	Vf	Wi	Ed	Ag

Ad – Accutediarrhoea; Ag – Amoebiasis and giardiasis; Ed – Eye disease; In – Influenza; Me – Measles; Mu – Mumps; Tb – Tuberculosis; Tf – Typhoid fever; Tg – Teeth and gum; Vf – Viral fever; Vh – Viral hepatitis; Wi – Worm infection.

Disease Zonation

A significant aspect of medical geography is to delineate the area under study into different disease zonation, which provides the foundation to disease regionalization. In present study the zonation of disease has been drawn on the basis of disease ranking and disease intensity.

Disease Zones

On the basis of disease ranking results, the Jammu region is divided into three first order and seven second order zones (Table 17.5) and depicted cartographically (Figure 17.6).

Table 17.5 Disease zones

S.No.	Zones		Name of zones
1	I order Ad zone	II order (a) Ad-In zone	Chenani, Mendhar, Sunderbani, Kalakote, Bishna, Samba, Hiranagar.
		(b) Ad-WI zone	Ramnagar, Rajouri, Thanamandi, Akhnoor
		(c) Ad-Ag zone	Jammu
		(d) Ad-Vd zone	GoolGulabGarh, Billawar
		(e) Ad-Tg zone	Gandoh, RanbirsinghPora
2	In Zone	(a) In-Ad zone	Ramban, Kishtwar, Surankote, Budhal, Doda, Punch, Bashohli
		(b) In-Vf zone	Banihal
		(c) In-Tg zone	Reasi
3	Tg Zone	(a) Tg-Ad zone	Nowshehra, Udhampur, Kathua
		(b) Tg-In zone	Bhaderwah, Thathri.

Ad – Acute Diarrhoea, In – Influenza, Tg – Teeth and gum, WI – Worm Infection, Vf – Viral fever, Ag – Amoebiasis and Giardiasis.

Intensity Zones

In order to delineate disease intensity zones, rank coefficient method has been followed, which reveals the area where diseases occur more frequently. Greater the frequency greater will be the intensity. Considering the lowest value of disease intensity rate as one, each tehsil has been ranked for every selected disease. Further, disease ranking coefficient for each tehsil has been calculated by using Kendal formula:

$$Ri = \frac{Dr_1 + Dr_2 + Dr_3 + ...Dr_n}{N}$$

Ri = Ranking Coefficient of each tehsil
Dr_1, Dr_2, Dr_3 .... Ranks occupied by the tehsils for disease r_1, r_2, r_3 ... r_n
N = Total number of diseases taken into
The rank coefficients of all the 30 tehsils have been further divided into four groups by quartile method (Figure 17.7).

High Intensity Zone

The high intensity zone (R > 13.23) forms an area of Chenani, Bashohli, Bishna, Samba, Kishtwar, Reasi, Bhaderwah and Thathri tehsils. The area coincides with Ad – In (Chenani, Bishna, Samba), In-Ad (Kishtwar – Basholi), Tg – In (Bhaderwah – Thathri) and In–Tg (Reasi). Factors responsible for high intensity are unsafe drinking water, poverty, high percentage of rural population low agriculture productivity, high altitude low temperature, etc.

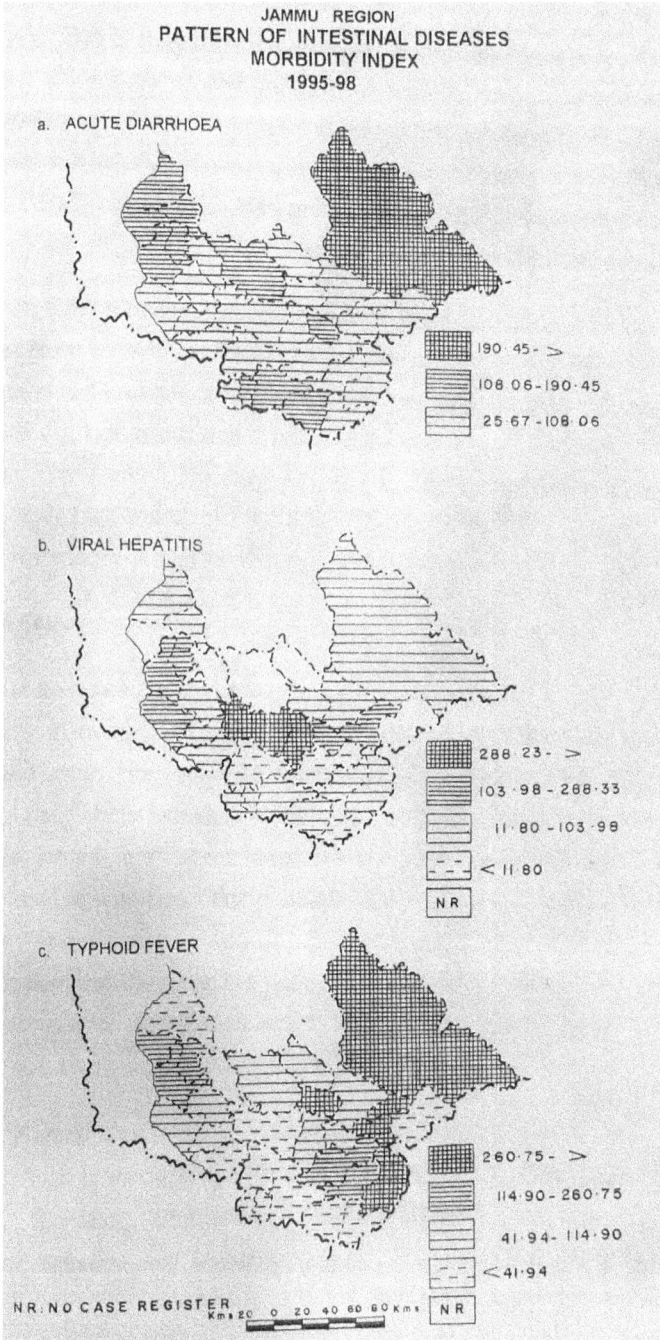

JAMMU REGION
PATTERN OF INTESTINAL DISEASES
MORBIDITY INDEX
1995-98

a. ACUTE DIARRHOEA

190·45- >
108·06-190·45
25·67-108·06

b. VIRAL HEPATITIS

288·23- >
103·98 - 288·33
11·80 -103·99
< 11·80
N R

c. TYPHOID FEVER

260·75- >
114·90- 260·75
41·94- 114·90
< 41·94
N R

NR: NO CASE REGISTER
Kms 20 0 20 40 60 80 Kms

Figure 17.6 Disease ranking 1988–98

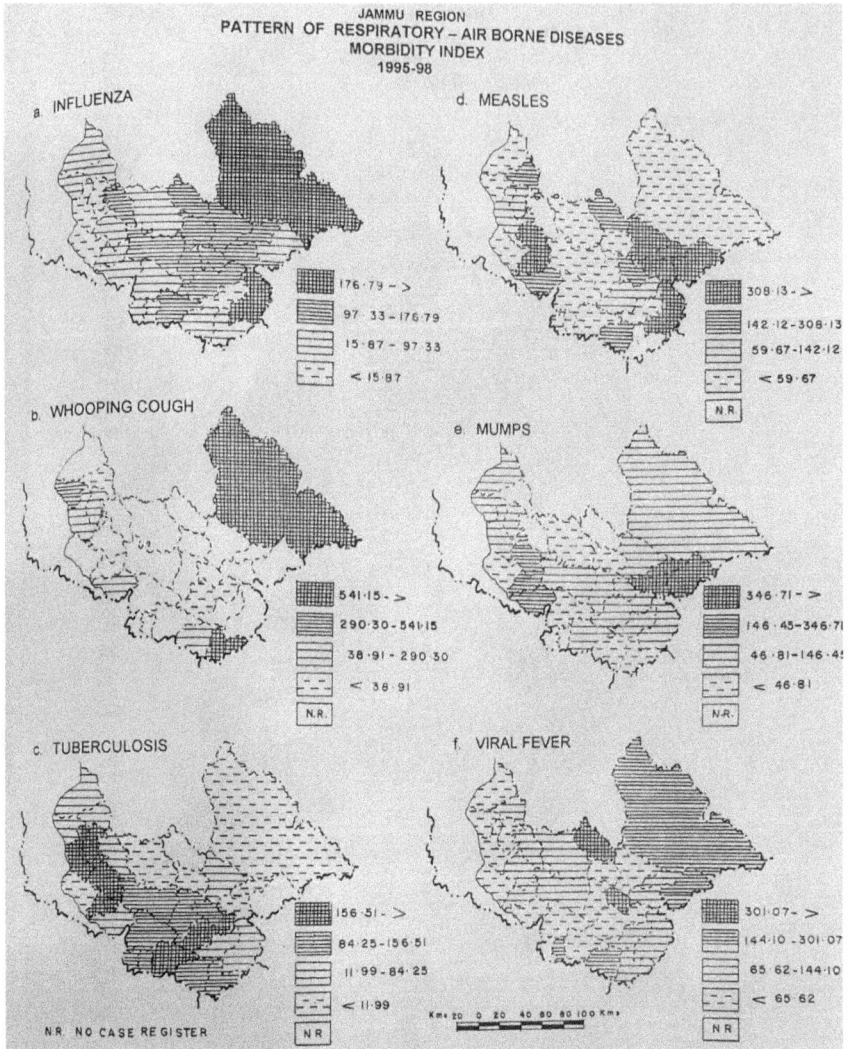

Figure 17.7 Disease zones

Moderately High Intensity Zone

This zone (R 11.86–13.23) covers maximum areas of the region. It comprises tehsils of Udhampur, Banihal, Mendhar, Surankote, Ramnagar, Sunderbani, Kalakote and Akhnoor. These tehsils coincide with zones of Ad-In (Sunderbani, Kalakote and Mendhar), Tg-Ad (Udhampur), Ad-Wi (Ramnagar and Akhnoor), In-Vf (Banihal) and In-Ad (Surankote). Unsafe drinking water, mountainous terrain, malnutrition, indoor air pollution, poverty, forest surroundings, low literacy, poor accessibility, high concentration of rural population are the major responsible factors.

Moderately Low Intensity Zone

The tehsils of Gandoh, Budhal, Doda, Rajouri, Thanamandi, Ramban, Hiranagar and Billawar exhibit pattern of moderately low intensity (R 9.46–11.86). The area coincides with In-Ad (Budhal, Doda and Ramban), Ad-WI (Rajouri and Thanamandi), and Ad-In(Hiranagar) zones.

Low Intensity Zone

The low intensity (R<9.46) is witnessed in Billawar, RanbirsinghPora, Nowshehra, Punch, GoolGulabGarh, Kathua and Jammu. These tehsils correspond with following zones Ad-Vf (Billawar, GoolGulabGarh), Ad-Tg (RanbirsinghPora), Tg-Ad (Kathua, Nowshera), In-Ad (Punch) and Ad-Ag (Jammu). The factors responsible for low intensity are proximity to major urban centres, good accessibility.

Thus, diseases accounting for high morbidity in Jammu region are mainly infectious and parasitic in nature, essentially associated with physical and socio-economic variables. Characteristics features of these health problems are that they often stem from poverty, unhygienic living conditions, malnutrition, unsafe drinking water, low literacy and poor health education.

Conclusion

With growing awareness about environment and ecology, the world over scientists, researchers and other academicians have been continuously emphasizing on the significance and importance of all related aspects. It was understood earlier also that diseases ailments and body disorder are basically the results of man's surroundings, i.e., physical, social, cultural and economic. Jammu region is a diverse area physiographically and climatically, characterized by lofty mountains, hills, valleys and plains. The region constitutes four physiographic divisions, viz, Foot Hills, plains, Outer Himalayas, Lesser Himalayas and Greater Himalayas. These physiographic divisions are crisscrossed by numerous rivers, streams, seasonal torrents and nullahs. The Jammu region experiences diverse climate conditions ranging from hot and humid to alpine type. Study area, demographically, socially, culturally and economically also shows a greats diversification.

The prevalence of communicable, deficiency and non-communicable disease throughout the region have affected large section of population and caused serious state of morbidity and poor health standards. In this regard, measures to control the intensity of disease need to be effectively implemented. The climate and physiographic have close interaction with various diseases in the region, in addition to various cultural environmental factors. In order to check morbidity, the various measures have to be taken into consideration, such as, to raise economic and female education levels, accessibility to health centres; to check malnutrition problems; to improve personal hygienic practices, and living conditions, to avoid modern living practices which creates stress and strain both mental and physical; avoiding heavy meals; smoking and excessive drinking; taking regular and moderate daily exercise, restraining from extra marital sexual relations, etc.

References

Agnihotri, R.C. (1995). *Geomedical Environment and Health Care: A Study of Bundelkhand Region*. Rawat Publications, Jaipur (India).

De Jayasree and Gollerkeri, R.S. (1984). Morbidity of infections hepatitis in Vadodara. *Annals of The National Association of Geographers India*, Vol. 4, No. 2, pp. 56–65.

Farooqi, M.Y. (1977). Diet and deficiency diseases in the Trans. Ropti Plain. *The Geographer*, Vol. 24, No. 1, pp. 43–55.

Hazra, J. (1987). Disease association in West Bengal, *Annals of the NAGI*, Vol. 7, No. 3, pp. 51–62.

Hughes, C.C. and Hunter, J.M. (1970). Disease and development in Africa. *Social Science and Medicine*, Vol. 3, No. 4, pp. 443–493.

Learmonth, A.T.A. (1984). Geography of health: A prologue. *The Indian Geographical Journal*, Vol. 59, No. 1, pp. 1–5.

May, J.M. (1950). Medical geography: Its methods and objectives. *The Geographical Review*, Vol. 40, pp. 9–41.

18 Food Insecurity in Rural Areas of Bandipore District, Jummu & Kashmir, India

Rais Akhtar

Introduction: Geo and Social Ecology of the Region

Bandipore district of Jammu and Kashmir state falls in the Western Himalayas that constitutes 25% of the total population of the Himalayas. It located on the banks of the Lake Wular which is said to be the largest fresh water lake in the whole Asia. This lake is the home to many migratory birds.

Nearly 50% of cultivated land is irrigated-paddy, maize and fodder are mainly grown. The average rainfall in the region has been recorded as 589 mm, with about 82 rainy days. The region covered varied topography plains, karewas, hill and wetlands and water bodies. The kind of topography in general makes the areas prone to flooding and drought conditions (Raina, 2002). Within the district, Bandipore tahsil has the least number of villages, as most of the area is under forest or water bodies. In most villages, the water is lifted by pumps and distributed through irrigation canals. The high content of clay soils becomes very loose when wet and hard when dry. Women and young ones are mainly engaged in the production of handicrafts such as carpet making, embroidery and shawl making. The people mostly belong to low income group (Gov, 2001).

The situation is exacerbated by low productivity, unemployment and under-employment. The region lacks basic services including sanitation (Plate 1) and safe drinking water. Average literacy rate is 51% lower than the national average (59.5%). Most people in the region suffer from waterborne diseases and malnutrition.

Methodology

The study is based on four villages which have been selected in tahsil Bandipore and 25 households were selected from each village based on stratified random sampling in order to administered questionnaire containing questions from the standard questionnaire used for similar surveys in other mountainous areas. The questions pertain to purchase of food items by location, response of people when faced with negative shocks, particularly the rising prices of food items, and major key drivers of food insecurity in the study area. The study is basically based of primary data obtained through the questionnaire.

DOI: 10.4324/9781003329459-18

Purchase of Food Items by Location

Figure 18.1 shows the distribution of locations from where people of the region purchased rice. Rice is a staple crop and constitutes major part of people diet. The region supply only about half of its requirements. Some 22% is obtained from other village and 16% from the town.

Potatoes are heavily consumed in the region and the region met only about 44% of its requirements and the remaining come from another village, town and outside mountain (Figure 18.2).

Rice Purchased by Location

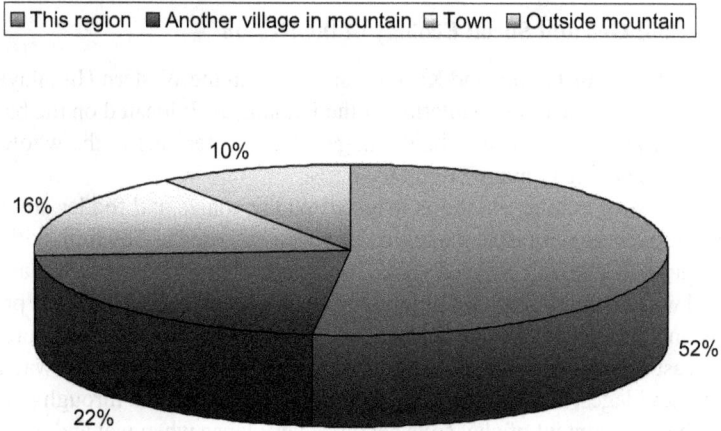

Figure 18.1 Rice purchased by location.

Potatoes Purchased by Location

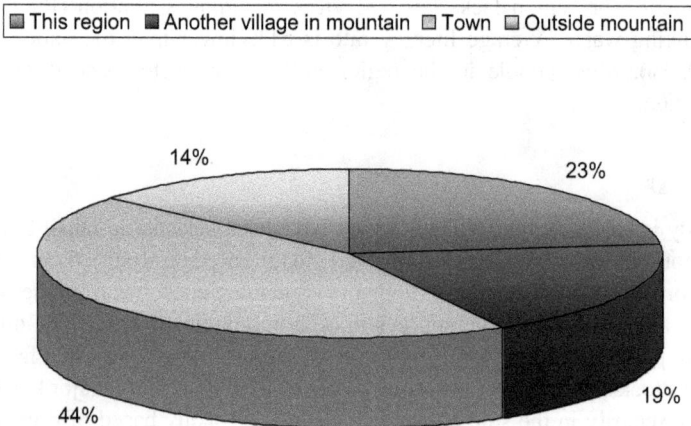

Figure 18.2 Potatoes purchased by location.

Onion are utilized in the preparation of different food items 32% of its requirements is met purchased from outside mountain followed with 26% from the region and another 26% from another village in the mountain (Figure 18.3).

Even vegetables like tomatoes the region is not self-sufficient with about 32% supply is obtained from outside mountain and 20% and 19% from another village and town respectively. The region production of tomatoes met is only 29% of the requirements of the study area (Figure 18.4).

Onion Purchased by Location

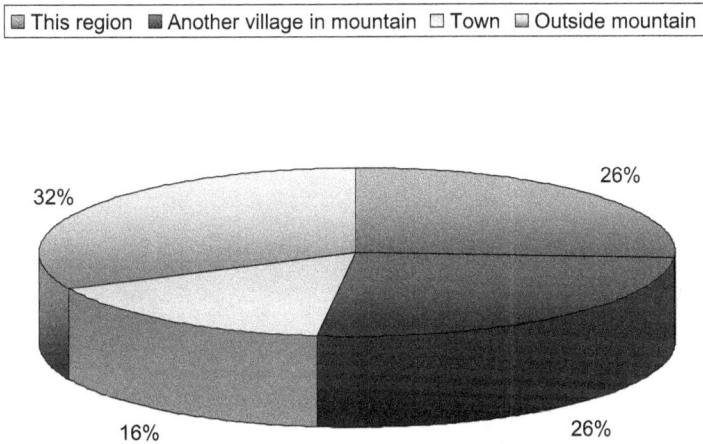

Figure 18.3 Onion purchased by location.

Tomatos Purchased by Location

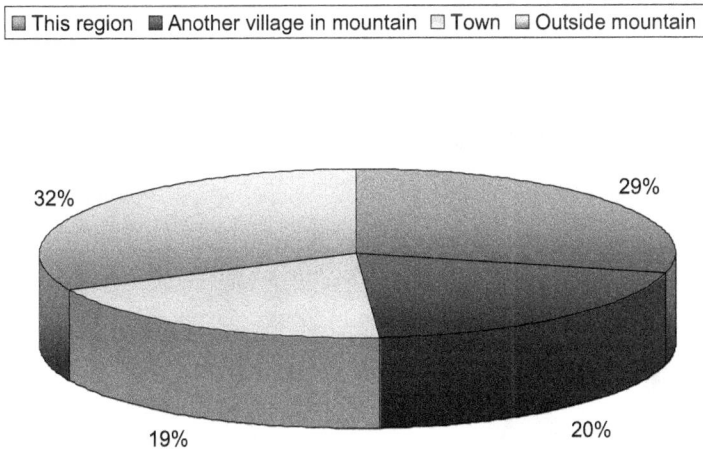

Figure 18.4 Tomatoes purchased by location.

For poultry people are heavily dependent on the nearby town, i.e., Bandipore. The region produces only 19% of the requirements and about 21% obtained from other village (Figure 18.5).

However, regarding the availability of milk the region produces nearly 83% of its requirements the remaining comes from another village (13%), town (3%) and outside mountain (1%) (Figure 18.6).

Chicken Purchased by Location

▨ This region   ■ Another village in mountain   ▢ Town   ▨ Outside mountain

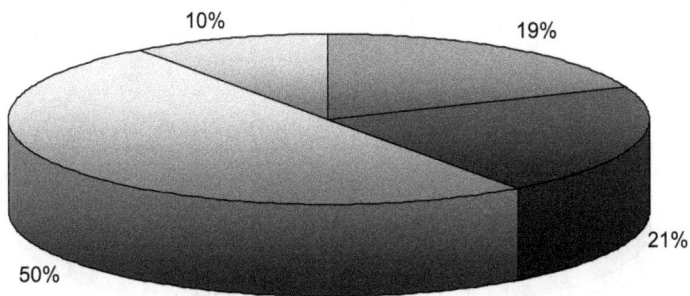

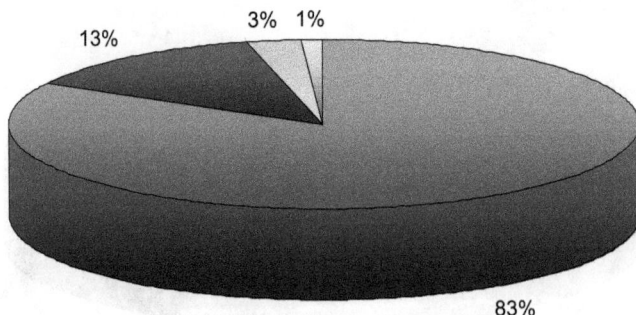

Figure 18.5 Chicken purchased by location.

Milk Purchased by Location

▨ This region   ■ Another village in mountain   ▢ Town   ▨ Outside mountain

Figure 18.6 Milk purchased by location.

Response of households when faced with negative shocks (Figure 18.7)
Key drivers in food insecurity (Figures 18.8 and 18.9):

Response of Households when faced with Negative Shocks

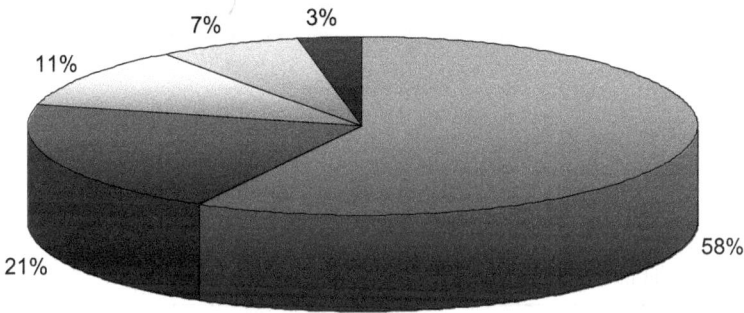

Draw down on savings   Borrow Money
Obtain assistance from family members   Obtain assistance from outside
Seek new way to earn income

7%   3%
11%
58%
21%

Figure 18.7 Response of households when faced with negative shocks.

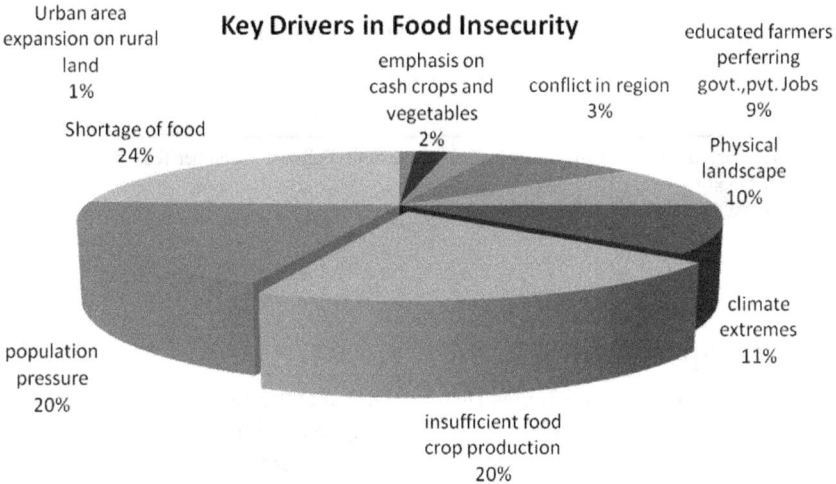

Key Drivers in Food Insecurity

Urban area expansion on rural land 1%
emphasis on cash crops and vegetables 2%
conflict in region 3%
educated farmers perferring govt.,pvt. Jobs 9%
Shortage of food 24%
Physical landscape 10%
climate extremes 11%
population pressure 20%
insufficient food crop production 20%

Figure 18.8 Key drivers in food insecurity.

Temperature Rise in last 20 years

■ High   ■ Medium   ▨ Low

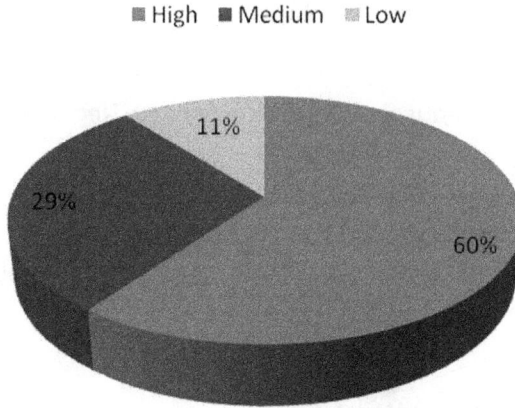

Figure 18.9 Temperature rise in last 20 years.

Table 18.1 Diet intake in the village (per head) in Chattibandy

	Chapattis,	Egg, no.	Butter, gm	Rice, gm	Vegetables, gm	Meat, gm	Curd, gm	Fruits, gm	Sugar, gm	Oil Fat, gm
Average of the day	3.07	1.00	58.33	216.90	206.70	100.00	110.10	30.00	40.00	40.00
Average of the week	21.48	7.00	350.58	1518.30	1446.90	700.00	770.70	210.0	280.00	280.00

Note: One chapatti is made of 70 grams of wheat/maize flour; Eggs are in number Tea – number of cups, other items in grams.
Source: Based on field work in the village.

Diet and Nutrition Intake

However, it is evident from Tables 18.1 and 18.2, that the diet intake and nutritional availability (k.cal.) is satisfactory when compared with recommended dietary allowances by the Indian Council of Medical Research (Gopalan et al., 1999). Though people suffer from malnutrition due to lack of nutrients rich foods (milk, meat, fruits and vegetables), the analysis is based on a village-Chattibandy, one of the selected villages.

Table 18.2 Nutrition intake (K. Cal.) in the village (per head) in Chattibandy

	Chapattis	*Eggs*	*Butter*	*Rice*	*Vegetables*	*Meat*	*Curd*	*Fruits*	*Sugar*	*Oil Fat*	*Total*
Average of the day	716.00	86.00	380.00	851.00	26.00	194.00	66.00	20.00	168.00	360.00	2867

Note: Recommended dietary allowances (K. cal.) for a person with moderate work is – 2,545.
Based on *Nutritive Value of Indian Foods*, by C. Gopalan et al., ICMR, National Institute of Nutrition, Hyderabad, 1999.

Conclusion and Suggestions

The study reveals that agricultural land does not produce enough for the population. As a result of this there are serious shortages of food items in the region. The analysis shows that most of the food items are purchased from outside the villages under study. Nearly 50% poultry comes from outside mountains (300–500 kms).

Nearly 58% people draw money from savings when faced with negative shocks, and only 21% borrow money. Major drivers related to food insecurity in the region are shortage of food (44%) population pressure (20%), extreme weather events (11%) and physical landscape (10%). However, the diet survey carried out in one of the four villages does show that at least in terms of K.calories majority of the population is better off, though they suffer from diseases as a result of malnutrition.

Thus the study reveals that households in general are not food-insecure and food security can be achieved also through purchase of food in such scenario, where land is not able to support the people, there is a need to promote non-agricultural activities in the rural areas in the mountains to enable people to increase their food purchasing power.

In order to make agriculture sustainable rain water harvesting and revival of traditional and new water storage system are crucial for water storage but must be adapted to the more extreme weather events. There is a need to restrict conversion of food crop land (rice) to orchards.

In order to make farmers self-sufficient in food production with objective food security and better livelihood, Jammu and Kashmir Government encourages farmers to use high yielding variety of seeds, providing lift irrigation pumps and foot spray pumps on 50% subsidy basis farmers are being encouraged to grow saffron and mushroom cultivation particularly in tribal belts. Agricultural land is being covered under land development programme, i.e. land levelling and bench terracing. Animal husbandry, fisheries and horticulture are being encouraged by the government (J&K, 2001). Besides, there is need to promote non-agricultural activities in agriculturally low productive areas with objective to enhance food purchasing capacity of the population. These include poultry and fish farming, sericulture and bee farming, and promoting handicrafts.

Acknowledgement

I am grateful to the field investigator – Dr. Ravi Kumar Bhat and his team members from Bandipore for conducting questionnaire survey in the four selected villages in Bandipore Tahsil. I am also thankful to him for conducting diet survey in the study area.

References

Gopalan, C et al. (1999) *Nutritive Value of Indian Foods*, Indian Council of Medical Research, Hyderabad.
Government of India. (2001) *District Census Handbook Baramulla, Jammu & Kashmir*, Registrar General's Office, New Delhi.
J&K. (2001) *Jammu & Kashmir Development Report* Chapter 3, Socio-Economic and Administrative Development, Srinagar, J&K.
Raina, A.N. (2002) *Geography of Jammu & Kashmir State*, Radha Krishan Anand & Co., Jammu.

19 Coronavirus Outbreaks in Union Territories of J&K and Ladakh

Rais Akhtar and Arshad Ahmad Lone

Introduction

No study has been carried out on the environmental and regional aspects of Coronavirus (COVID-19) infection and its distribution. Its origin and diffusion from the epicentre in China reveals a strange pattern. From geographical distribution it looks in the beginning a curious pattern emerged, that major COVID-19 hotpots – Wuhan (China), Iran, Northern Italy, France, Spain, U.K. and the United States, are located in temperate cooler regions, and rarely in warm tropical areas. For instance, the temperature in these regions ranges between 12 and 15 degrees Celsius during late March and early April 2020 (Akhtar, 2021)

Swedish geographer Torsten Hagerstrand propounded his theory of types of disease diffusions, and the current trends in COVID-19 outbreaks reflects the Expansion and Re-location diffusions, the virus expanded from foci areas., to surrounding China and Hong Kong ,and re-located as far off areas in Iran, Italy particularly northern Italy (Milan, Bologna and Venice) which are closer to Alps mountains. France, Spain and the United Kingdom are other European countries suffered from COVID-19 disaster. This hypothesis seems to have defeated when COVID-19 impacted Australia and Brazil and other south American countries. Undoubtedly, the United States suffered heavily and ranks number one both in terms of morbidity and mortality. Thus it is evident and the causes of disease distribution varies regionally. These regional variations in disease distribution is an important component in the study of medical geography. The discipline of medical geography focuses on understanding spatial patterns of health and disease as related to the natural and social environment. Conventionally, there are two primary areas of research within medical geography: The first deals with the spatial distribution and determinants of morbidity and mortality centring on regional factors. While the second deals with health care patterns and planning, health-seeking behaviour and the provision of health services, identification and minimizing inequalities in the region (Akhtar, 1991a, 1991b).

COVID-19 and Climate

It is evident from the diffusion of COVID-19 that the cooler climate is congenial for its spread (Figure 19.1) opposite to malaria plasmodium which preferred

DOI: 10.4324/9781003329459-19

warmer climate, and enhanced warming causing occurrence of altitudinal rise of malaria in highlands and sub-mountainous areas.

In another empirical research on climate ad COVID-19, the MIT scientists concluded that, the total number of cases in countries with mean temperature greater than 18 degree Celsius and absolute humidity more than 9 g/m^3 in January–February–early March is less than 6%. (MIT, 2020). "Within the US, the study reveals that the outbreak also shows a north-south divide. Northern (cooler) states have much higher growth rates compared to southern (warmer) states. The spread of 2019-nCoV has been limited in Texas, New Mexico and Arizona," they reported in the study (MIT, 2020). Another study suggests that warmer or more humid climates do not slow the spread of the COVID-19, according to a study published in *Science* by researchers from the Princeton University (Basu, 2020; Collins, 2020). A study published in the *Nature* entitled 'Why COVID outbreaks look set to worsen this winter', especially in regions that don't have the virus spread under control (Mallapaty, 2020).

At the same time some studies carried out in early and the middle of 2020 takes into consideration the COVID-19 cases in Tibet and high-altitude regions of Bolivia and Ecuador in comparison to the low lying regions. It suggested that the population in Bolivia, Ecuador and Tibet living above 3,000 meters (9,842 feet) reported significantly lower levels of confirmed infections than their lowland counterparts.

EurAsian Times further explained that "Just three populations in the world have been found to have genetic adaptations to altitude: Himalayans, Ethiopian highlanders and Andeans. This is why the coronavirus is exploding on Peru's Pacific coast, particularly Lima, where most residents descend from Andean ancestors, while the country's mountain communities are thus far not greatly affected by the virus (Eurasiantimes, 2020).

COVID-19 in J&K and Ladakh UTs

In Jammu & Kashmir union territory, two suspected cases with high virus load were detected and isolated on 4 March in Government Medical College, Jammu. One of them became the first confirmed positive case on 9 March 2020. Both individuals had a travel history to Iran.

As of 30the December 2020, total number of positive cases in Jammu and Kashmir were 121,217 and 1,880 deaths. In view of the pandemic, the government has declared Lakhanpur containment zone on the NH-44 with a buffer of 500 meter radius and Jawahar tunnel area, on either side as red zone while Jammu division's Kishtwar district has been kept in the category of green zone unlike rest of the districts of the division (G.K., 2020).

Total number of cases reported in Jammu and Kashmir union territory is 108,150. A total of 67,933 cases were reported from Kashmir division and 46,840 cases were reported from Jammu division. Urban centres of these two divisions, i.e., Srinagar and Jammu reported highest number of cases 23,974 and 21,853 respectively. The reason for height cases in these two cities could be because of height travel history and mass movement of people from other locations. Least number of cases where reported from Shopian, Ramban, Poonch and Samba. The above results showed urban locations and commercial areas are high-prevalence zones in comparison to rural areas (Table 19.1).

Total number of cases reported in Jammu and Kashmir union territory were 108,150. About 67,933 cases were reported from Kashmir division and 46,840

Table 19.1 District-wise Distribution of Coronavirus Cases in Jammu and Kashmir UT December 2020

S.No	District	Number of cases
1	**Srinagar**	**23,974**
2	Baramulla	7,674
3	Budgam	7,219
4	Kupwara	5,322
5	Pulwama	5,249
6	Anantnag	4,646
7	Bandipora	4,513
8	Ganderbal	4,316
9	Kulgam	2,623
10	Shopian	2,397
11	**Jammu**	**21,853**
12	Udhampur	3,896
13	Rajouri	3,659
14	Doda	3,264
15	Kathua	2,954
16	Kishtwar	2,630
17	Samba	2,600
18	Poonach	2,415
19	Ramban	2,017
20	Reasi	1,552
Total		**114,773**

Source: Directorate of Health Services, Jammu & Kashmir.
Note: As on 30 December 2020, total number of cases in Jammu & Kashmir UT increased to **1,21,217.**

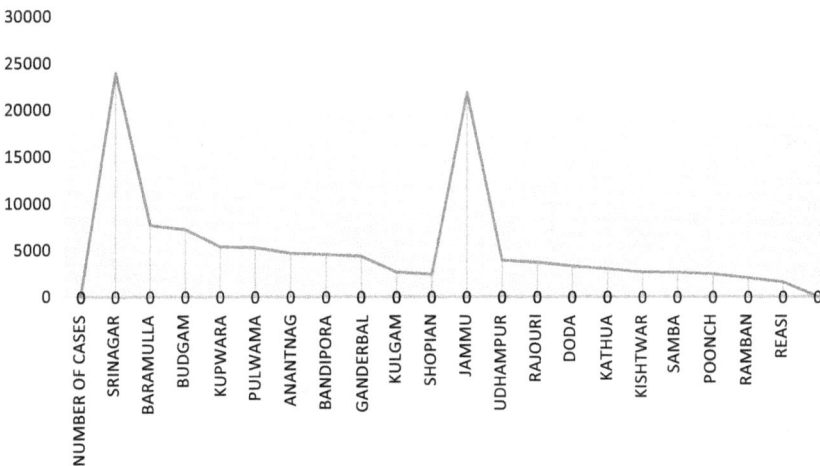

Figure 19.1 Line graph of distribution of COVID-19 cases in Jammu and Kashmir Union Territory (drawn by the authors, based on Govt. Data).

cases were reported from Jammu division. Urban centres of these two divisions, i.e., Srinagar and Jammu reported highest number of cases, i.e. 23,974 and 21,853 respectively. The reason for high cases in these two cities could be because of high travel history and mass movement of people from other locations. Least number of cases were reported from Shopian, Ramban, Poonch and Samba. Thus the data reveals that urban locations and commercial areas marked as high-prevalence zones in comparison to rural areas (Figure 19.2 and Table 19.2).

Total number of cases reported from union territory of Ladakh were 9,155. Leh district reported higher number cases, i.e. 7,074 in comparison to Kargil with 2,081 cases. Some 977 of those infected with COVID-19 in Kargil were returnees from Iran in comparison of only 87 cases in Leh. Besides those infected Kargil 60 cases were foreign returnees in comparison to 472 in Leh.

It is evident from the distribution pattern that there is marked uneven distribution with urban areas like Srinagar , Jammu and Leh reporting higher number cases in comparison to districts like Shopian , Ramban and Kargil. The uneven distribution could be because of locational, economic and functional nature of areas. Kishtwar is the only district in Jammu division that was categorized in green zone. A report reveals, that barring Kishtwar, all Jammu & Kashmir districts classified as orange zones (Greater Kashmir, 2021a).The distribution pattern provides a good opportunity for researches to delve into locational, socio-economic, cultural and behavioural determinants of spatial pattern of coronavirus distribution.

Table 19.2 District-wise distribution of COVID-19 cases in Ladakh Union Territory

District	Kargil	Leh	Total
Cases	2,081	7,074	9,155

Source: Ministry of Health and Family Welfare, Government of Ladakh.

Note: As on 30 December 2020, the total number of cases increased to 9,447.

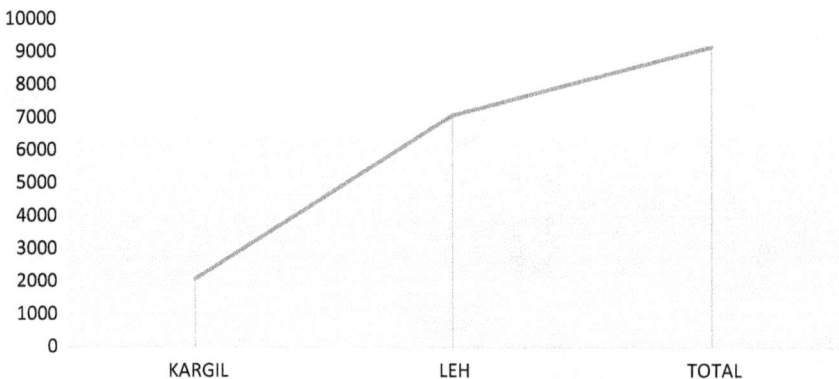

Figure 19.2 District-wise distribution of COVID-19 cases in Ladakh Union Territory, December 2020.

Changing COVID-19 Scenario after Second Wave

Due to the impact of Second wave from March 2021 onwards, there has been an enormous increase in the incidence and deaths due to COVID-19. Following Table 19.3 and Figure 19.3 depict the spatial pattern of COVID-19 cases by district in both J&K and Ladakh Union territories.

Table 19.3 COVID-19 Incidence in Jammu & Kashmir (31 August 2021)

S.No	District	Cases
1	**Srinagar**	**71,992**
2	Baramulla	23,816
3	Budgam	23,067
4	Kupwara	15,285
5	Pulwama	14,210
6	Anantnag	16,439
7	Bandipora	9,604
8	Ganderbal	10,073
9	Kulgam	11,402
10	Shopian	5,593
11	**Jammu**	**52,974**
12	Udhampur	11,375
13	Rajouri	11,078
14	Doda	7,505
15	Kathua	9,276
16	Kishtwar	7,134
17	Samba	4,745
18	Poonach	6,308
19	Ramban	5,984
20	Reasi	6,435
Total		**324,295**

Source of Data – Directorate of Health Srevices Kashmir (August 2021).

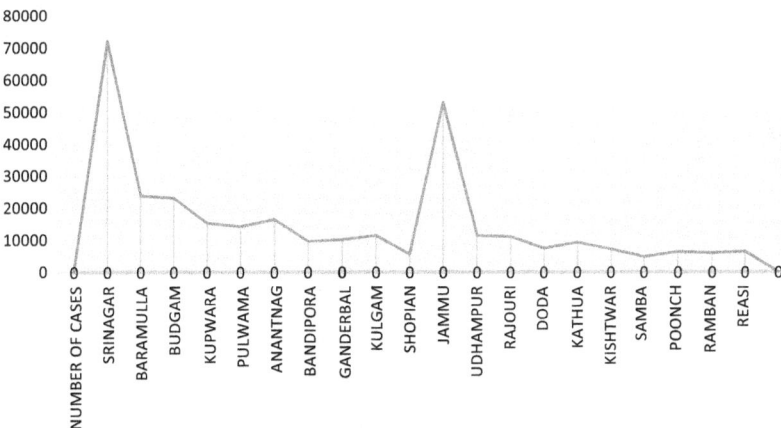

Figure 19.3 Line graph showing COVID-19 cases distribution in Jammu and Kashmir Union Territory.

Total number of cases reported in Jammu and Kashmir union territory was 324,295 with 201,481 cases were reported from Kashmir division and 122,814 cases belonged to Jammu division. Again urban centres of these two divisions, i.e., Srinagar and Jammu reported highest number of cases i.e. 23,974 and 21,853 respectively. The reason for higher incidence in these two large cities could be because of high population density and mass movement of people from other locations. Lowest number of cases were reported from Shopian, Kishtwar, Poonch and Samba. Thus the above pattern showed that high-density urban locations and commercial hubs are high-prevalence zones in comparison to smaller urban rural areas (Figure 19.4 and Table 19.4).

Total number of cases reported from union territory of Ladakh was 26,468. Leh district, an important tourist destination, particularly popular among foreigners, reported 20,447 cases in comparison to 6,021 cases in Kargil.

Thus the number of cases reported showed uneven distribution with areas like Srinagar, Jammu and Leh reporting more cases and districts like Shopian, Kishtwar and Kargil reporting lower number of cases. The uneven distribution of COVID-19 cases may be associated with the population pattern and because of locational, economic and functional nature of these cities. According to a report, Delta variant was likely responsible for 85% cases in Jammu & Kashmir (G.K., 2021) COVID-19 infections is still a serious health problem. In late 2021, Srinagar municipality had to impose curfew for 10 Days in Five Areas, and some areas in Baramulla town (G.K., November, 2021b). Report in January 2022 states that ten new Omicron cases detected in Jammu & Kashmir UT, including five cases in Jammu region.

Table 19.4 District-wise distribution of COVID-19 cases in Ladakh Union Territory

District	Kargil	Leh	Total
Cases	6021	20447	26468

Source: Ministry of Health and Family Welfare (August 2021).

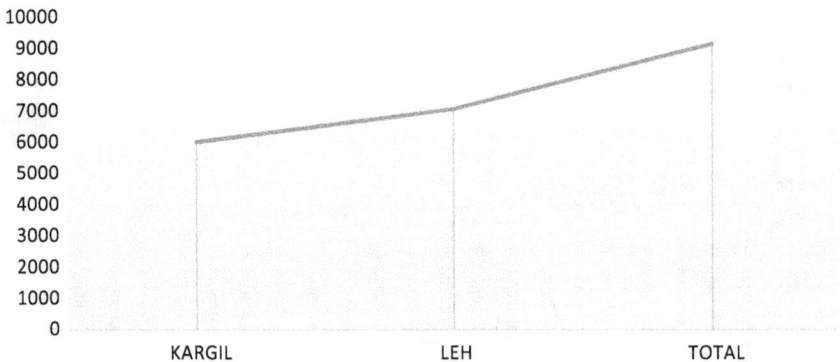

Figure 19.4 District-wise distribution of COVID-19 cases in Ladakh Union Territory, August 2021.

Ladakh suspends winter tourist activities due to Omicron threat (Daily Excelsior, 2022; Times of India, 2022) (Figures 19.5 and 19.6; Tables 19.5 and 19.6).

Table 19.5 Covid-19 Incidence in Jammu & Kashmir Union territory (30 March, 2022)

S.No	District	Cases
1	**SRINAGAR**	**102,252**
2	BARAMULLA	37,588
3	BUDGAM	34,803
4	KUPWARA	18,416
5	PULWAMA	21,717
6	ANANTNAG	22,419
7	BANDIPORA	13,674
8	GANDERBAL	13,626
9	KULGAM	16,313
10	SHOPIAN	6,304
11	**JAMMU**	**74,654**
12	UDHAMPUR	14,613
13	RAJOURI	12,547
14	DODA	11,284
15	KATHUA	11,232
16	KISHTWAR	8,956
17	SAMBA	5,826
18	POONACH	7,621
19	RAMBAN	8,050
20	REASI	8,650
	TOTAL	**447,923**

Source: Directorate of Health Services, Jammu & Kashmir.
Note: As on 30 March 2022, total number of cases in Jammu & Kashmir UT increased to 447,923.

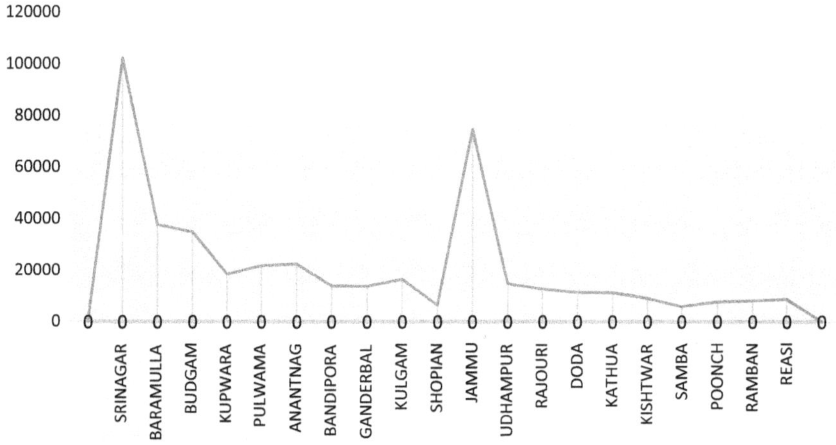

Figure 19.5 District-wise distribution of COVID-19 in Jammu & Kashmir Union Territory.

Source: Directorate of Health Services, Jammu & Kashmir.
Note: As on 30 March 2022, total number of cases in Jammu & Kashmir UT increased to 447,923.

Table 19.6 District-wise distribution of COVID-19 cases in Ladakh Union Territory, 30 March 2022

District	Kargil	Leh	Total
Cases	8,031	28,225	36,256

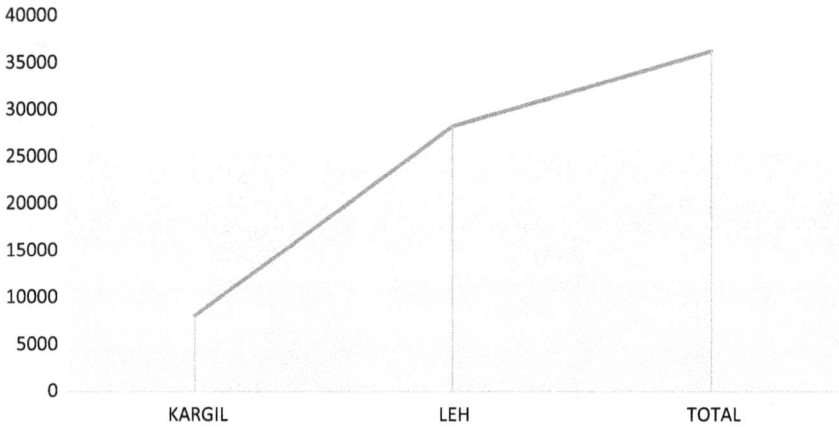

Figure 19.6 District-wise distribution of COVID-19 cases in Ladakh Union Territory, (30 March 2022).

Source: Ministry of Health and Family Welfare, Government of Ladakh.

Note: As on March 2022, total number of cases increased to 36,256.

COVID-19 figures for Jammu & Kashmir, for the three different periods of time (2020, 2021 and 2022) indicate that Srinagar shows the highest number of COVID-19 cases followed by Jammu. In Ladakh, Leh records the highest number of COVID-19 cases followed by Kargil. As explained population density, size of urban areas, high connectivity between rural and smaller urban areas and the popular destinations for tourism, particularly among foreign tourists in Ladakh region, may be explained in the distribution pattern of COVID-19 in Jammu & Kashmir and Ladakh union Territories.

Conclusion & Suggestions

It is evident from the discussion that COVID-19 originally occurred in a cooler climate which was initially considered congenial for its spread in some geographical regions including mountainous areas. The opposite is true of malaria plasmodium which prefers a warmer climate, and enhanced warming causing occurrence of altitudinal rise of malaria in highlands and sub-mountainous areas. At the same time current pattern of COVID-19 shows that it also occurred in varied climate zones including the one which experience greater precipitation. The studies also suggest that the links between COVID-19 cases and temperature are less certain. Some studies reported link between temperature and COVID-19, while others have not. However, higher temperatures are associated with a lower number of cases in Turkey, Mexico, Brazil and the U.S. (Guite, 2020). There is need to conduct regional studies with focus on geoecology of COVID-19 in both developed and developing regions of the union territories of Jammu & Kashmir and Ladakh which encompasses Greater Himalaya, Lower Himalaya and plain areas

Research must also focus on the significance of socio-geographical and human behavioural determinants in the outbreak COVID-19 and its diffusion particularly the impact of residents from Ladakh, returning from Iran and from other foreign countries. Attempt should also be made to study the impact of this pandemic on mental stress of the population.

With such devastating experience, we will be able to make better sense of COVID-19 in terms of how we can learn from this pandemic experience-identify our weaknesses in handling regional health crisis by strengthen the health infrastructure. We can utilize the strategies adopted successfully in other regions and to ensure all communities are adequately cared for, and reducing disparities, and have access to the health care resources including accessibility of COVID-19 vaccines particularly in remote hilly areas, and to discourage vaccine hesitancy, that will contribute to the overall wellbeing of our society. With efforts made by the Govt. of India, and Jammu & Kashmir and Ladakh Governments, and increased awareness of people who adopted COVID appropriate behaviour, will considerable decline in COVID-19 cases. Despite these efforts, COVID-19 cases are still rising in both Jammu & Kashmir and Ladakh. As on 14 August 2022, there were 473,782 cases with 4,777 deaths in Jammu & Kashmir, and 28,271 cases with 228 deaths in Ladakh. Nonetheless the people of the UTs of Jammu & Kashmir and Ladakh must adopt COVID appropriate behaviour, and utilize facility of booster vaccine to control the spread of COVID-19 in the region.

Acknowledgement

We are thankful to the concerned health directorates of J&K and Ladakh Union Territories for providing necessary data.

References

Akhtar, R. (1991a) *Environment and Health: Themes in Medical Geography*, Ashish Publications, New Delhi.

Akhtar, R. (1991b) *Health Care Patterns and Planning in Developing Countries*, Greenwood Press, New York.

Akhtar, R. (2021) *Coronavirus(COVID-19) Outbreaks*, Environment and Human Behaviour, Springer.

Basu, M. (2020) Warmer Climate Does Not Prevent Covid-19 Spread, Large Number of People Still Vulnerable, *The Print*, 19 May.

Collins, F. (2020) Will Warm Weather Slow Spread of Novel Coronavirus? June 2.

Daily Excelsior. (2022) 10 new Omicron Cases Detected in Jammu & Kashmir UT including 5 Cases in Jammu Region, January 12.

Eurasiantimes. (2020) Coronavirus In Tibet: Limited Impact Of COVID-19 In High Altitude Regions Like Tibet? June 3 (https://eurasiantimes.com/coronavirus-in-tibet-scientists-exploring-links-between-covid-19-high-altitude-regions-like-tibet)

G.K. (2020) Barring Kishtwar, All J&K Districts Classified as Orange Zones, *Greater Kashmir*, Srinagar, October 9.

G.K. (2021a) Delta Variant Responsible for 85% Cases in J&K, *Greater Kashmir,* August 22.

G.K. (2021b) *Srinagar Imposes Curfew for 10 Days in Five Areas, Greater Kashmir, Srinagar*, November 10.

Guite, H. (2020) How Does Weather affect COVID-19?, *Medical News Today*, August 16.

Mallapaty, S. (2020) Why Covid Outbreaks Look Set To Worsen this Winter, *Nature*, Vol. 586, October 29, p. 653.

MIT. (2020) Warm, Humid Climate Linked To Slower COVID-19 Transmission, MIT Study, March 26.

Times of India. (2022) *Ladakh Suspends Winter Tourist Activities due to Omicron Threat*, January 4.

20 Levels of Malnutrition by Altitude Based on Body Mass Index (BMI) among School Children (14–18 years) in Gujjar Community of Greater Kashmir Himalayas

G. M. Rather

Introduction

Body Mass Index (BMI), calculated as weight in kilograms divided by the square of height in meters (kg/m^2). It is a standard measure of weight-for-height that is commonly used to classify malnutrition grades as: underweight, overweight, and obesity of people and correlate with future health risks of morbidity and death. Globally, one-fifth of the world's population accounting about 1.2 billion adolescents are under the grip of under nutrition with the number ever increasing mostly in developing countries, on the other hand USA, a developed country revealed that if obesity continues to increase at current rate nearly 90% of adults and two thirds of children by 2050 will be overweight or obese (Zargar et al., 1997). The state of Jammu and Kashmir where the present study was carried out also revealed dual burden of malnutrition in adults as well as children with 25% of women and 28% of men are too thin; and 17% of women and 6% of men are overweight or obese as compared to children where 35% are stunted, 15% are wasted and 26% are underweight (NFHS-3, 2006). The BMI of an individual is often the result of many inter-related factors which can be classified as immediate, underlying and basic, complex, ranging from biological and social to environmental factors, political instability and slow economic growth, to highly specific ones such as the frequency of infectious diseases and the lack of education (WHO, 2014). Moreover suitable aspects of natural environment, or example, mineral traces in water, the geological nature of bed rock material and specific biologic complexes also affect human health and may lead to long-term chronic ailments (Armstrong, 1971).

Impact of nutrition on health of children is not a recent approach in Medical geography but has attracted the attention of experts for the last more than half a century and plenty of literature is available at national and international level but very less is available at regional level. Some notable contribution at regional level are as under.

Zargar found 46.39% of the Population in Pulwama and 52.32% in Anantnag affected by goiter (Zargar et al., 1997). Rather identified 29% children suffering from various deficiencies because of low birth weight as compared to ICMR (Rather, 2004). Mayer identified diverse agricultural activities responsible for highest prevalence of anaemia in Kashmir valley (Mayer, 2007). Dewan

DOI: 10.4324/9781003329459-20

attributed it to poor socio-economic status as the root causes of malnutrition with 25.2% women as compared to male's 20.2% in Punjab (Dewan, 2008) and Shukla revealed the encouraging association between literacy and malnutrition (Shukla, 2011). Akhtar and Koundal in Jammu & Kashmir while Krishnan in Tamil Naidu find out the regional disparity in health care patterns and planning process of the state mainly responsible for malnutrition (Akhtar, 2009; Koundal, 2012). Krishnan et al. (2012) and Khan et al. found feeding practices sub-standard before the recommended standards leading to parallel increase in the malnutrition with 14.1%, 17.2%, and 16.8% of the children in Jammu, Kashmir, and Ladakh (Khan & Khan, 2012). Gull Assessing the women health aspects of Gujjar and Bakerwal Community of Jammu & Kashmir found the health of Gujjar and Bakerwal women's very deteriorating the reasons being high family pressure as all the work is being done by women folk besides rearing of animals, illiteracy, lack of awareness about the schemes, and facilities through meager provided by the govt. agencies and lack of health services (Gull, 2014). The present research paper was an attempt to analyse BMI of school children in the age group of 14–18 years and variation of same with altitude. The study reveals the health status of sample children and shall be of great help for future health planning in this mountainous region.

Study Area

Greater Kashmir Himalayan range is one of the most important physiographic divisions of Jammu and Kashmir State and extends uninterruptedly for a length of 150 km from Sundran drainage basin of Anantnag in the south to Kazinag ridge of Baramulla in the north (Figure 20.1). Greater Kashmir Himalayan range is a massive topographical feature enclosing Kashmir Valley on the east–north east and north–northwest. The range lies between $33^0 22'32.02''$ N–$34^0 47'42.67''$ North latitude and $73^0 48'10.96''$ E–$75^0 34'22.23''$ East longitude. The mountainous range has an average altitude of 3,442 meters and stretches over an area of 8948.84 sq. Kms. The base contour of the range is around 1,800 meters in the south and gradually decreases to around 1,600 towards north. Below the base contour of the mountain range, the Valley of Kashmir has homogeneity in level. The region has a slope from 10^0 to 30^0 in the foothills and above 40^0 in the hilly areas. The present slope characteristics have evolved through a sequence of events including spectacular changes in base level through faulting, folding and the consequent rejuvenation of drainage channels with pronounced effects on land forms in general and slope in particular (Raza et al., 1978), The region is inhabited by Gujjar community with very low socio-economic development.

Data Base and Methodology

The present research work was based on both primary and secondary data. Large data both primary and secondary was collected and generated from different sources. A comprehensive methodology used for the present study and described under the following headings.

Delineation of Study Area and demarcation of Altitudinal Zones (Unit of Study)

Base map of the study area was delineated from 19 SOI Toposheets and processed digitally in GIS environment. Greater Kashmir Himalayan Range was divided into the following seven altitudinal zones with the help of software's like ERADAS Imagine 9.0 and Arc view GIS 3.2a (Table 20.1).

Selection of Sample Villages, Sample Households, and Sample Children (14–18 years)

Stratified Random Sampling technique was used for selection of around 20% of sample villages (60) and 20% of sample households (2,080) in proportion to total number of villages and households from each altitudinal zone. A Sample of 2,093 children, one male and one female, falling in 14–18 years were selected for Micro study (Table 20.2). Geo-coordinates and altitude of each sample village was measured with the help of GPS during field survey (Table 20.3).

Table 20.1 Altitudinal zones by area

Altitudinal zone	Alt. in meters amsl	Area in Sq. Kms.	Area in % to total area
A	1,600–1,750	499.18	5.59
B	1,750–1,900	510.22	5.70
C	1,900–2,050	490.19	5.47
D	2,050–2,200	516.45	5.70
E	2,200–2,350	515.38	5.75
F	2,350–2,500	530.12	5.96
G	2,500–6,000	5887.30	65.83
	Total	8948.84	100

Source: Compiled by Author.

Table 20.2 Sample frame of the study

Alt. zone	Alt. in mts. (AMSL)	Total area (Km²)	Revenue villages			Number of households			Number of children (14–18 years) for MICRO STUDY		
			Total in area	Sample	Percentage of sample	Total in sample villages	Sample	Percentage of sample	Male	Female	Total
A	1,600–1,750	499.18	9	2	22.22	460	92	20.00	45	46	91
B	1,750–1,900	510.22	31	6	19.35	1,000	200	20.00	103	103	206
C	1,900–2,050	490.19	71	14	19.71	2,380	476	20.00	242	241	483
D	2,050–2,200	516.45	72	14	19.44	2,290	458	20.00	228	228	456
E	2,200–2,350	515.38	81	16	19.75	2,790	558	20.00	275	275	550
F	2,350–2,500	530.12	40	8	20.00	1,480	296	20.00	153	154	307
	Total	8948.84	304	60	19.73	10,400	2,080	20.00	1046	1,047	2,093

Source: Computed from SOI toposheets and census of India.

Table 20.3 Sample villages with altitude and geo-coordinates

S. No.	Village name	Lat./Long	Altitude (mamsl)	S. No	Village name	Lat./Long	Altitude (mamsl)
1	Grand	33°40'43" N 75°15'20" E	1,830	31	Dardpora Gugerpati	34°25'43" N 74°42'16' E	2,250
2	Hard kichloo	33°50'45" N 75°16'40" E	2,390	32	Aragam Nagbal	34°22'31" N 74°40'58" E	2,060
3	Gujran Batkot	33°56'34" N 75°18'07" E	2,186	33	Chithi Bande chaliwan	34°22'46" N 74°41'13" E	2,290
4	Ishnad	33°52'08" N 75°18'04" E	2,268	34	Argam Halwadi	34°22'30" N 74°40'57" E	2,055
5	Hapatnar	33°48'17" N 75°21'15" E	2,520	35	Sumlar Gujjarpati	34°22'30" N 74°43'41" E	1,885
6	Salia	33°55'28" N 75°17'26" E	2,210	36	Chuntimula Gujjarpati	34°24'23" N 74°44'05." E	1,980
7	Gous	33°52'09" N 75°18'32" E	2,190	37	Chatibandhi Gorhajan	34°23'40" N 74°42'25" E	1,835
8	Shojan	33°51'14" N 75°18'25" E	1,890	38	Malangam Gujjarpati	34°26'12" N 74°33'26. E	1,950
9	Grandwan	33°52'43" N 75°17'54" E	2,020	39	Mulkalama Gujjarpati	34°24'03" N 74°43'34" E	2,375
10	Lidu	33°57'31" N 75°18'52" E	2,049	40	Gujjarpati Muqam	34°26'58" N 74°34'36" E	2,250
11	Rishkobal	33°08'03" N 75°17'51" E	2,350	41	Kudara	34°25'03" N 74°47'01" E	2,410
12	Nagbal	33°52'32" N 75°20'25" E	2,260	42	Dachna Gujjarpati	34°26'02" N 74°30'56" E	1,680
13	Dragund	34°25'51" N 75°04'55" E	2,120	43	Manobal	34°30'15" N 74°30'15" E	2,055
14	Narasthan	34°13'27" N 75°05'25" E	2,250	44	Londa	34°18'24" N 74°10'20" E	2,010

(Continued)

Table 20.3 (Continued)

S. No.	Village name	Lat./Long	Altitude (mamsl)	S. No	Village name	Lat./Long	Altitude (mamsl)
15	Guturu	34°30'27" N 75°25'20" E	2,160	45	Nilzab	34°30'25" N 74°12'42" E	2,290
16	Hajannar	34°04'31" N 75°03'37" E	1,893	46	Potwari	34°19'45" N 74°12'20" E	2,065
17	Nogh	33°55'46" N 75°11'10" E	2,142	47	Khaitan	34°30'50" N 74°30'35" E	1,935
18	Bangidar	33°54'40" N 75°14'09" E	2,354	48	Nowgam	34°28'19" N 74°14'25" E	1,980
19	Basmia	33°55'44" N 75°11'06" E	2,262	49	Lahkoot	34°21'45" N 74°20'52" E	1,955
20	Faqir Gujri	34°24'16" N 74°38'50" E	2,089	50	Rashiwari	34°40'55" N 74°48'45" E	2,410
21	Shal khud	34°10'59" N 74°54'58" E	2,215	51	Shiltra	34°19'14" N 74°12'08" E	1,835
22	Nagbal Gujjarpati	34°15'22" N 74°34'25" E	1,967	52	Inderdaji	34°20'12" N 74°08'54" E	1,950
23	Khanan	34°18'47" N 74°51'59" E	2,030	53	Khuri payeen	34°39'55" N 74°45'30" E	2,250
24	Poshkar	34°14'26" N 74°58'05" E	2,080	54	Khuri Bala	34°42'15" N 74°45'40" E	2,315
25	Pahalnar	34°20'49" N 75°51'59" E	2,142	55	Wadur bala	34°18'26" N 74°11'06" E	2,058
26	Wangat	34°19'33" N 75°06'50" E	2,195	56	Turkkpora	34°32'52" N 74°26'35" E	2,386
27	Astan mohla	34°15'29" N 74°54'44" E	2,048	57	Wanpur	34°28'12" N 74°16'30" E	2,036
28	Yarmukam	34°17'44" N 74°47'11" E	2,360	58	Wahalutar	34°46'22" N 74°14'32" E	2,253
29	Tsunt Wali war	34°47'14" N 74°54'28" E	2,370	59	Potus	34°45'20" N 74°12'28" E	2,146
30	Waniarm	34°17'44" N 74°48'30" E	2,295	60	Naidhu	34°25'23" N 74°16'55" E	1,684

Source: Based on GPS readings during Sample survey- 2018.

Table 20.4 WHO classification of malnutrition grades based on BMI

Classification		$BMI(kg/m^2)$
	Principal cut-off Points	**Additional cut-off points**
Underweight	<18.50	<18.50
Severe thinness	<16.00	<16.00
Moderate thinness	16.00–16.99	16.00–16.99
Mild thinness	17.00–18.49	17.00–18.49
Normal range	**18.50–24.99**	**18.50–22.99**
		23.00–24.99
Overweight	≥25.00	≥25.00
Pre-obese	25.00–29.99	25.00–27.49
		27.50–29.99
Obese	≥30.00	≥30.00

Source: WHO, 2007.

Anthropometric Measurements

Anthropometric measurement of 2,093 sample children, comprising of one male and one female from each household of sample village was carried out to obtain data regarding weight and height. BMI was calculated from weight and height for both male and female children of all altitudinal zones. As physical dimensions of body are influenced by nutrition particularly during the rapidly growing period of early adulthood thus body measurement can also provide information regarding malnutrition so height, weight, and BMI was measured. The weight and height of all individual sample children were measured by using digital weight measuring machine and non-stretchable tape. The method was repeated more than once and the mean was of the reading was taken as final.

BMI of all the sampled children was calculated by employing the formula

$$BMI = \frac{\text{Weight in Kg.}}{\text{Height in m}^2}$$

WHO Classification of 2007 (Table 20.4) was used to classify the sample children into different grades of malnutrition based on BMI

Results and Discussions

Weight for Age

Analysis of the Table 20.5 reveals that average weight in the age group 14–18 years was 22.105 kg for males and only 21.270 kg for females. It was very interesting to note that the average weight of both male and female children was very less than the ICMR recommended weight for children of different age groups. Weight of children varies from one sample village to another. There was a decline in the weight of both male and female children with the increase in the altitude (Figure 20.2). Largest differences in the calculated weight than the ICMR recommended weight was noted and the reason could be large nutrition need for the fast growth on one side and very less attention of parents towards child because of being engaged in primary activity of collection of fire wood from the forest and herding of animals, getting of water from large distances, etc.

Height for Age

Analysis of the Table 20.5 reveals that average height in the age group 14–18 years was 123.3 cm for males and only 119.6 cm for females. It was very interesting to note that the average height of both male and female children was very less than the ICMR recommended height for children of different age groups. Height of children varies from one sample village to another. There was a decline in the height of both male and female children with the increase in the altitude (Figure 20.3).

Body Mass Index

Analysis of the Table 20.5 reveals that average BMI in the age group 14–18 years was 14.53 for males and only 16.96 for females. It was very interesting to note that

Table 20.5 Average weight for age, average height for age and average body mass index among sample children (14–18 years)

Altitudinal zone with alt. in meters (amsl)	No. of male sample children (male & female – same ratio)	Average anthropometric values for parameters among sample children by sex					
		Average weight (Kg)		Average height (cms.)		Average body mass index	
		Male (ICMR standard : 33.10 Kg)	Female (ICMR standard: 34.00 Kg.)	Male (ICMR standard: 140 cms)	Female (ICMR standard: 139 cms.)	Male (ICMR standard: 18.5)	Female (ICMR standard: 18.5)
Zone-A (1,600–1,750)	91	20.130	20.730	122.8	120.1	14.58	13.35
Zone-B (1,750–1,900)	206	24.962	23.467	123.7	123.6	15.52	15.31
Zone-C (1,900–2,050)	483	23.380	22.340	126.0	119.6	15.97	15.04
Zone-D (2,050–2,200)	456	21.930	21.350	123.3	119.8	15.08	14.68
Zone-E (2,200–2,350)	550	21.720	20.190	121.8	117.6	16.15	15.04
Zone-F (2,350–2,500)	307	20.510	19.540	122.0	117.0	14.44	13.77
Total	2,093	22.105	21.270	123.3	11 119.6	15.96	14.53

Source: Sample Survey-2018.

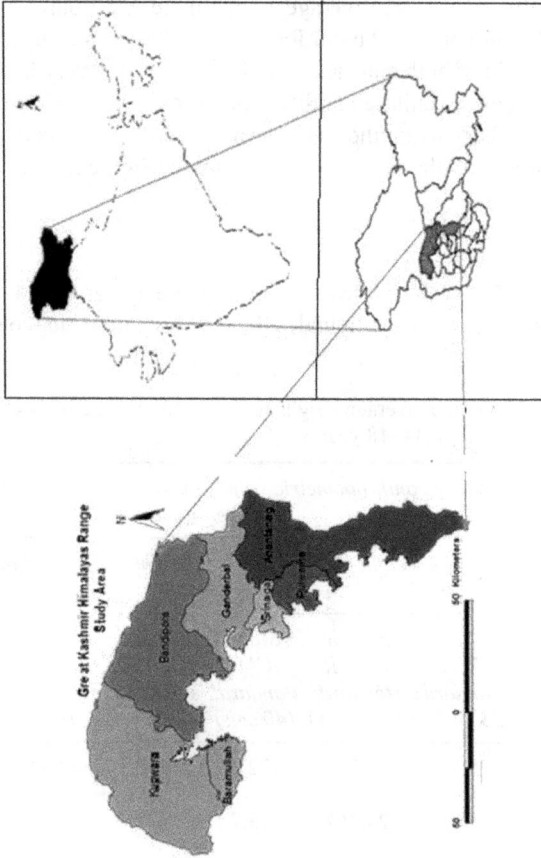

Figure 20.1 Average weight for age among sample children (14–18 years).

Source: Based on Sample Survey – 2018.

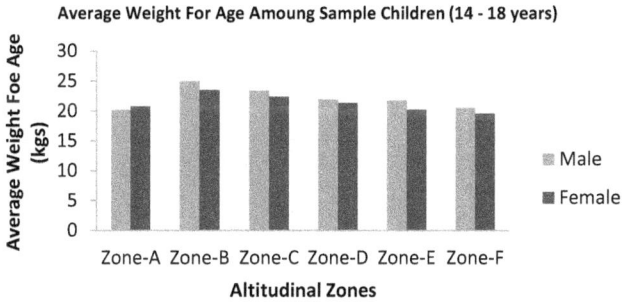

Figure 20.2 Average height for age among sample children (14–18 years).
Source: Based on Sample Survey- 2018.

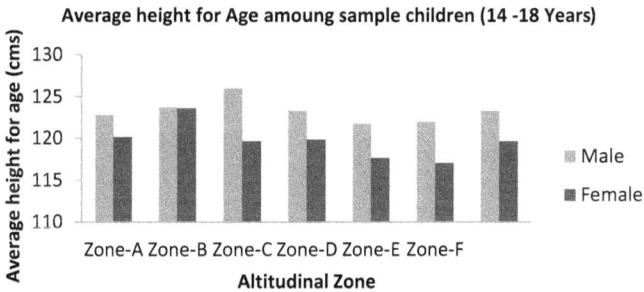

Figure 20.3 Average BMI among sample children (14–18 years).
Source: Based on Sample Survey – 2018.

the average BMI of both male and female children was very less than the ICMR recommended weight for children of different age groups. BMI of children varies with altitude (Figure 20.4)

Grades of Malnutrition among Sample Children (14–18 Years) Based on BMI

Analysis of the Table 20.6 reveals very good variation of malnutrition grades with altitude. Percentage of male sample children with BMI >18 was 10% in altitudinal zone A and increased with altitude to 21% in altitude zone F. Severe/Moderate Grades of malnutrition among male children increased from 4% in altitudinal zone A to 15% in altitudinal zone F and grade of Obese was only 3% in altitudinal zone A and no Obese in altitudinal zone E and F. While as in case of female sample children percentage was lower than males in each Grade of malnutrition with altitudinal zone (Figure 20.5).

Table 20.6 Variation in malnutrition grades (percentage) among sample children using BMI

Zone	Total BMI < 18		Severe/Moderate		Obese	
	Male	*Female*	*Male*	*Female*	*Male*	*Female*
A (1,600–1,750)	10.00	10.20	4.00	5.63	3.00	2.51
B (1,750–1,900)	15.60	16.14	9.03	10.15	2.00	1.60
C (1,900–2,050)	16.52	17.00	10.18	12.10	1.50	1.15
D (2,050–2,200)	17.34	17.65	11.58	12.14	1.20	1.00
E (2,200–2,350)	18.06	19.18	14.00	14.20	0.01	0.00
F (2,350–2,500)	21.00	21.60	15.00	15.10	0.00	0.00
Average	16.42	16.96	10.63	11.55	2.57	1.04

Source: Based on Sample Survey - 2018.

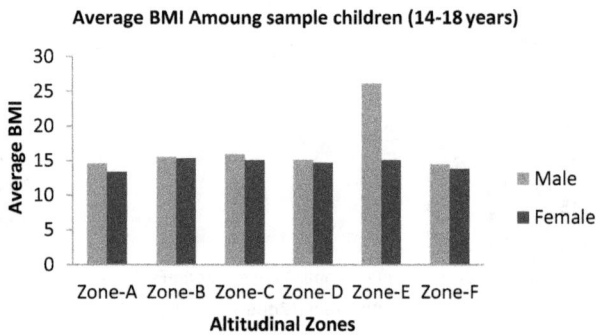

Average BMI Amoung sample children (14-18 years)

Figure 20.4 Variation in malnutrition grades (percentage) among sample children using BMI.

Source: Based on Sample Survey-2018.

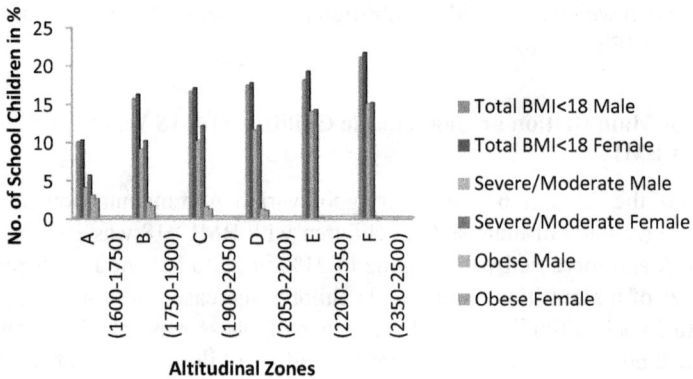

Figure 20.5 Altitudinal Zones.

Conclusion and Suggestions

The Study leads to the conclusion that mean height, weight, and BMI were 123.3 cm, 22.105 kg, and 15.96 kg/m² for male and 119.6 cm, 21.270 kg, and 14.53 kg/m² for female respectively and were less than ICMR standards. Near about 16.42% male and 16.96% female were having BMI <18.5 kg/m² while as 10.63% of male and 11.55 % of female were having severe-to-moderate malnutrition. Analyses reveals decrease of BMI with increase in Altitude. It may be attributed to traditional cropping practices and harsh agro climatic conditions which prevail there, besides non-availability crop essentials in the vicinity. The sample population face extended period of illness because of poverty and non-ability of Medicare and if present that too is of low standard thus affects the BMI. Some suggestions are recommended as under,

1 The present study shows a dismal picture of Malnutrition by analysing the BMI. So, Awareness Programmes/counselling, regarding Balanced diet, health, and physical activity are needed badly. These programmes should be encouraged through ICDS centres and Government schools.
2 Implementation of various schemes of NRHM at gross root level as only AHSA component of JSA scheme is found visible in the study area. Availability/ Appointment of Dietitian in every govt. and semi govt. institutions should be made under NHRM.
3 There is a need for development of agriculture and Horticulture in the area that too on scientific lines.

Acknowledgement

The author is highly thankful to ICSSR for providing funding assistance to our project "Assessment of Malnutrition among Children – A Micro level study of Gujjars of Greater Kashmir Himalayan Range". The present research paper is a part of the Project.

References

Akhtar, R., (2009). *Regional Planning for Health Care System in Jammu & Kashmir*, Concept Publishing Company, New Delhi, pp. 29–30.

Armstrong, R.W., (1971), Medical Geography and Its Geologic Substrate, In: H.L. Cannon and H.C. Hopps (eds.), *Environmental Geo-chemistry in Health and Diseases*, Boulder, CO, pp. 211–219.

Dewan, M., (2008). Malnutrition in Women, *Studies on Home and Community Science*, Vol. 2, Issue 1, pp. 7–10.

Gull, S., (2014). Assessment and Understanding of Gujjar and Bakerwal Women's Health in Jammu and Kashmir, *Journal of Business Management & Social Sciences Research*, Vol. 3, Issue 3, pp. 37–43.

Khan, Y., & Khan, N., (2012). Nutritional Status of Children (0-24 Months) in Jammu Kashmir and Ladakh Regions, *International Journal of Scientific and Research Publications*, Vol. 2, Issue 6, pp. 1–7.

Koundal, V., (2012). Poverty among Nomadic Gujjars – A Case Study of J & K And H.P, *International Journal of Marketing, Financial Services & Management Research,* Vol. 1, Issue 8, pp. 206–230.

Krishnan, M., Rajalakshmi, P. V., & Kalaiselvi, K., (2012). A Study of Protein Energy Malnutrition in the School Girls of a Rural Population, *International Journal of Nutrition, Pharmacology, Neurological Diseases,* Vol. 2, Issue 2, p. 142.

Mayer, I. A., (2007). Regional Analysis of Diet and Nutritional Anemia in Kashmir Valley, *Kashmir Journal of Social Science*, Vol. 2, pp. 45–50.

National Family and Health survey (NFHS-3), IIPS (2005–2006).

Rather, G. M., (2004), Levels of Mal-Nutrition in Preschool Children of Four Rural Communities in Bandipora and Gurez, *Geographical Review*, Geographical Society of India, University of Calcutta, Vol. 66, Issue 1, pp. 28–40.

Raza, M. et al., (1978), *The Valley of Kashmir- A Geographical Interpretation*, Vol. I- The Lands, Vikas Publications, New Delhi, pp. 15–45.

Shukla, B., (2011). Study to Assess the Nutritional Status of Under Five Children in a selected Sub Center of Bikaner, Rajasthan – India, *International Research Journal, Vol.* 1, Issue 1, pp. 1–10.

WHO, (2007), World Health Report, Geneva.

WHO, (2014). *Technical Report on Malnutrition*, World Health Organization, Geneva.

Zargar, A. H., Shah, J. A., Laway, B. A., (1997). Epidemiology of Goitre in School Children in Rural Kashmir (Pulwama District), *Journal International Medical Sciences Academy,* Vol. 10, issue 1, pp. 13–14.

21 Comparative Profile of COVID-19 Mortality in Jammu versus Kashmir Provinces in Union Territory of Jammu and Kashmir

Rajiv K Gupta, Shashi Sudhan Sharma, Richa Mahajan and Talat Jabeen

Introduction

COVID-19, a viral disease, emerged at the end of December 2019 and spread globally in a very short span of time. World Health Organization (WHO) declared COVID-19 as a pandemic on 11 March 2020 (ECDC 2020). Over 180 countries have been affected by COVID-19, resulting enormous morbidity and mortality worldwide (Djaharuddin et al. 2021 p.530–32). COVID-19 is caused by Severe Acute Respiratory Syndrome Coronavirus 2 (SARS-CoV-2), which belong to the family of Betacoronavirus genus (Zhu et al. 2020 p.727–33). Although the clinical presentation and symptoms of COVID-19 are similar to that of Middle East Respiratory Syndrome (MERS) and Severe Acute Respiratory Syndrome (SARS), the rate of spread is greater (Peeri et al. 2020 p.717–26). The first case of COVID-19 in India was reported on 30 January 2020 from Kerala, among Indian medical student who had returned from Wuhan, the epicentre of the pandemic (Andrews et al. 2020 p.490–92). Slowly, the pandemic spread to various states and union territories including Jammu and Kashmir. Two suspected cases with high virus load were detected and isolated on 4 March 2020 in Government Medical College, Jammu. One of them became the first confirmed positive case on 9 March 2020. Both individuals had a travel history to Iran. The worldwide traffic of people facilitated the rapid spread of COVID-19 reaching countries around all the continents in a very short span of time.

COVID-19 pandemic has caused lot of misery all over the world with millions of hospitalizations and deaths. Besides direct impact of the disease itself, indirect harms have arisen because of overwhelming of health systems, lockdown policies and economic struggle.

India is among the worst affected nations in the ongoing COVID-19 pandemic with over five lakh deaths by 31 January 2020, while UT of Jammu and Kashmir has reported 4,681 deaths till 31 January 2022. Since detection of SARS-CoV-2 in December 2019, in Wuhan city of Hubei province in China, lot of research has been carried out on epidemiological, demographic, clinical features, diagnostic and treatment aspects of the disease. It was followed by research on morbidity and mortality aspects of the disease which included variables like patient's age, sex or association of co-morbid conditions with increased risk of SARS-CoV-2 induced

DOI: 10.4324/9781003329459-21

adverse clinical outcome including mortality. Some studies have reported that COVID-19 was higher in males than females (Chen et al. 2020 p.507–13; Huang et al. 2020 p.497–506; Yang et al. 2020 p.475–81) while others do not show similar findings (Qian et al. 2020; Xu et al. 2020 p.1275–80). As cases evolved globally, it was noted that persons with underlying chronic illnesses are more likely to contract the virus and become severely ill.

Mortality rates provide an opportunity to identify and act on the health system intervention for preventing deaths (Lavergne and McGrail 2014 p.8–9). COVID-19 mortality is an important estimate to know the disease burden in the community (ECDC 2020).

COVID-19 has been a global threat that has already pushed the healthcare to its limits. Despite availability of vaccine since last more than one year, the pandemic is far from over. Due to SARS-CoV-2 being a relatively new virus, the data available is limited. The lack of data can impact decision-making for public health officials, thereby affecting the preventive measures that need to be taken to intercept this deadly pandemic. During review of literature, it was found that not much literature is available on mortality profile of COVID-19 patients, chiefly in UT of Jammu and Kashmir. This study aims to bring into light the comparative mortality profile of COVID-19 patients in Jammu and Kashmir provinces of UT of Jammu and Kashmir.

Material and Methods

Since the onset of COVID-19 pandemic in India, data on new cases, recovered cases and deaths are being calculated meticulously. In this context, COVID-19 dashboards have been set up at National level and State/UT levels.

In the UT of Jammu and Kashmir, State Surveillance Office (SSO) of both the provinces viz. Jammu and Kashmir have been diligently collecting the data regarding COVID-19 since the onset of the pandemic. The mortality data of COVID-19 was procured from both the provinces and analysed on the basis of gender, age groups, districts, presence or absence of co-morbidity and death since time of admission to the hospital. The data after analysis was duly represented by bar and pie diagrams.

Results

It was found that total deaths in the UT of Jammu and Kashmir since the onset of COVID-19 pandemic till 31 January 2022 was 4,681. Further analysis revealed that Kashmir division constituted 51% of these deaths viz. 2391 (Figure 21.1). Sex-wise distribution of deaths revealed that males were the worst sufferers with 62–63% of deaths being in males in both the provinces of UT of Jammu and Kashmir (Figure 21.2). Age-wise mortality analysis elicited that majority of deaths were reported in 50–69 years age group with Kashmir province reporting 46% in comparison to 43% of Jammu province in this age group. It was followed by deaths in >70 years age group while only upto 1% mortality was reported in <18 years age group (Figure 21.3). District-wise mortality in Jammu province revealed that

Jammu district was at the top with 1208 deaths and it was followed by 245 deaths in Rajouri district (Figure 21.4). In Kashmir province, Srinagar district reported 909 deaths, followed by Baramulla district at 301 deaths while Shopian reported least number of deaths at 60 (Figure 21.4). When mortality data was analysed on the basis of deaths per million population, the results revealed districts of Jammu and Srinagar were having maximum numbers viz. 790 and 735 deaths per million population, respectively. Reasi district in Jammu province reported only 140 deaths per million population (Figure 21.5). As far as co-morbid status of deceased patients was concerned, 79% of reported deaths had co-morbidity in Kashmir province as compared to 56% in the Jammu province (Figures 21.6 and 21.7).

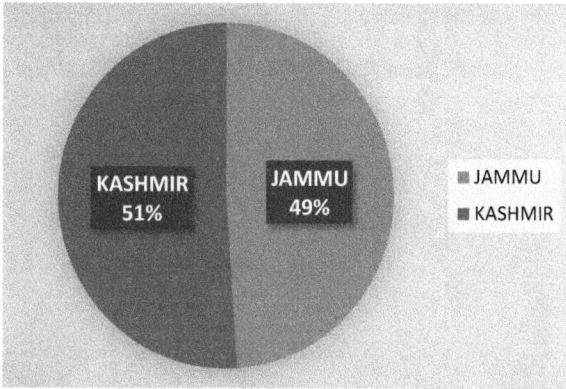

Figure 21.1 Deaths due to COVID-19 in Jammu and Kashmir.

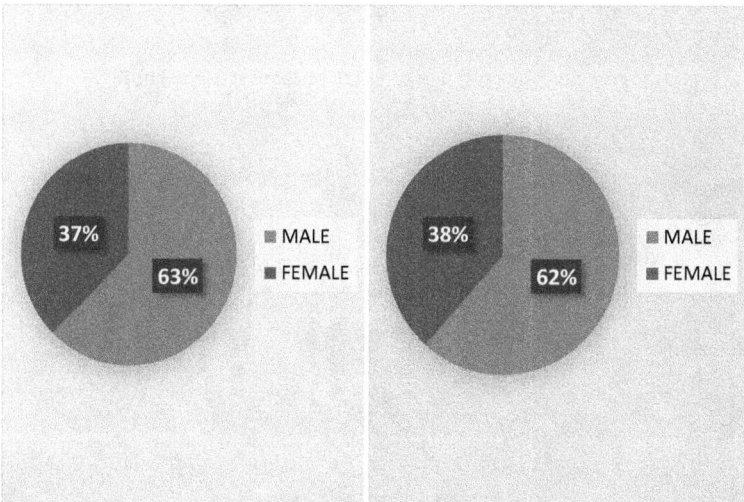

Figure 21.2 Sex-wise distribution of COVID-19 deaths in Jammu (left) and Kashmir (right) provinces.

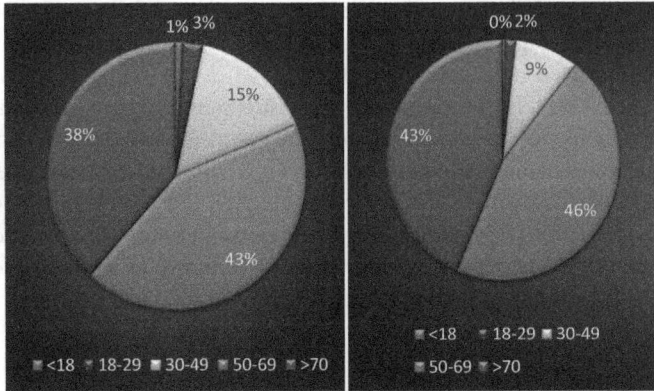

Figure 21.3 Age-wise distribution of COVID deaths in Jammu (left) and Kashmir (right) province.

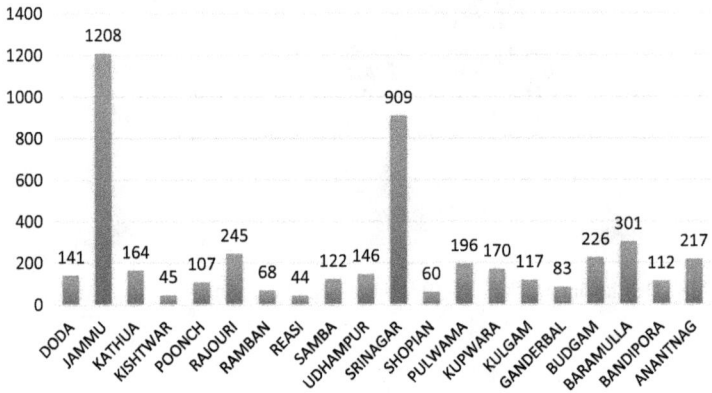

Figure 21.4 District-wise number of deaths in UT of Jammu & Kashmir.

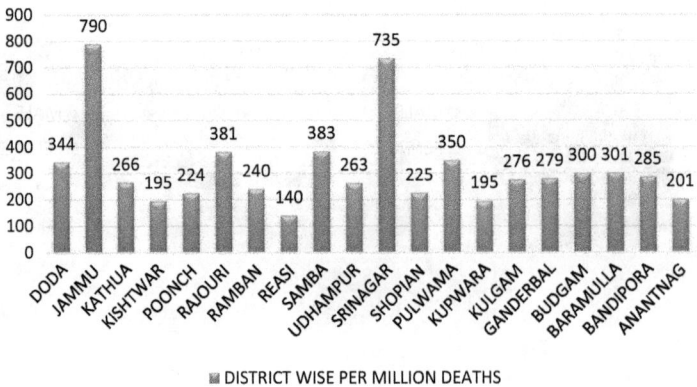

Figure 21.5 District-wise mortality profile of Jammu and Kashmir provinces (deaths per million population).

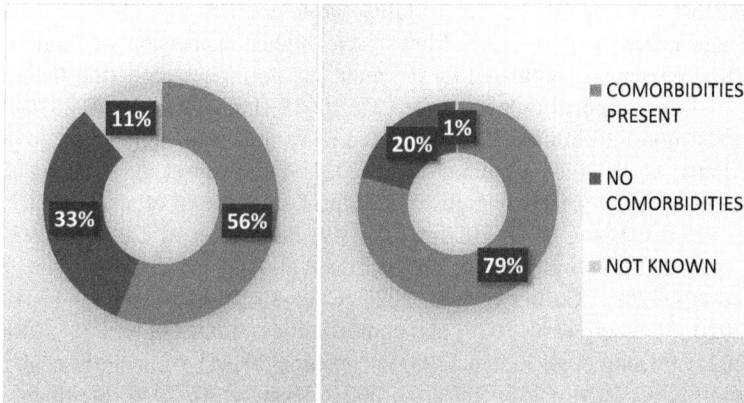

Figure 21.6 Co-morbid status of deceased COVID patients in Jammu (left) and Kashmir (right) provinces.

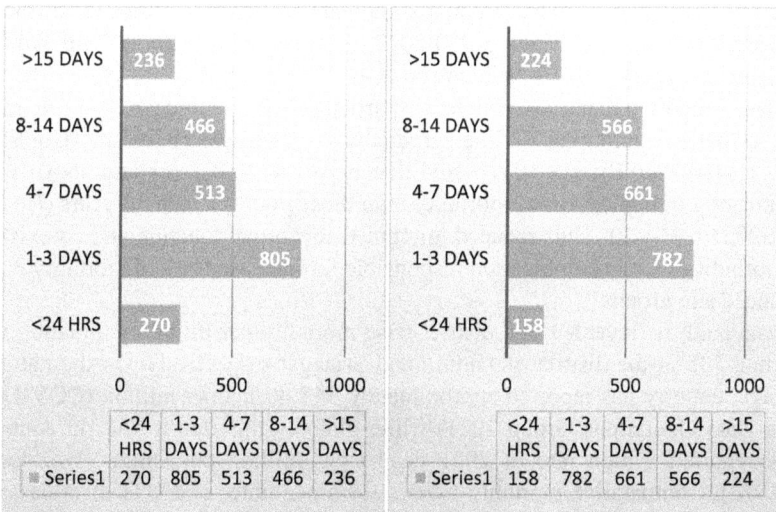

Figure 21.7 Duration of hospital stay of COVID-19 patients before death in Jammu (left) and Kashmir (right) provinces.

Discussion

COVID-19 pandemic has caused a large scale of morbidity and mortality in the entire world. The pandemic with its protean manifestation remains an unpredictable debacle, the spectrum of presentation varying from asymptomatic infection to a fulminant systemic inflammatory syndrome unleashed by the cytokine storm. The results of mortality in UT Jammu & Kashmir have revealed that male deaths were to the tune of two-third of total deaths and 50–69 years of age group was the worst affected. Further it was found that the patients aged 70 years and above were the second most severely affected in mortality, while only 1% mortality was reported in the age group of <18 years. Males have higher expression of angiotensin-converting enzymes-2 regulated by the male sex hormones rendering them more at risk and poor clinical outcomes for COVID-19 (La Vignera et al. 2020). Risk factors for mortality in COVID-19 reported in various studies included advanced age (Ahmad et al. 2021; Chauhan et al. 2021; Galloway et al. 2020 p.282–88; Ji Dong et al. 2020 p.1393–99; Joshi 2020 p.11–12; Knight et al. 2020; Li et al. 2020 p.110–18; Liang et al. 2020 p.1081–89; Salunke et al. 2020 p.1495–501; Shang et al. 2020; Steinberg et al. 2020; Zhou et al. 2020 p.1054–62), male gender (Ahmad et al. 2021; Galloway et al. 2020 p.282–88; Joshi 2020 p.11–12, Knight et al. 2020; Steinberg et al. 2020) and comorbidities (Ahmad et al. 2021; Chauhan et al. 2021; Ji Dong et al. 2020 p.1393–99; Joshi 2020 p.11–12; Knight et al. 2020; Nakakubo et al. 2020 p.858; Rai et al. 2021; Shang et al. 2020; Steinberg et al. 2020; Varghese et al. 2020 p.401–10; Zhou et al. 2020 p.1054–62) like diabetes mellitus, obesity, systemic hypertension, renal diseases, coronary artery disease (Chauhan et al. 2021) and malignancy. Most countries with available data have shown higher infection, mortality and fatality rates in males than in females though these rates vary from country to country and between regions (Bhopal and Bhopal 2020 p.532–33; Global health 50/50 2022; Pradhan and Olsson 2020 p.474–81; Scully et al. 2020 p.442–47; Williamson et al. 2020 p.430–36).

Leng J reported that older people are particularly prone to develop more infections as natural immunity declines gradually at older ages (Leng and Goldstein 2010 p.1120–24). Biswas M reported that those aged 50 years and above were significantly at higher risk of mortality than those younger than 50 years (Biswas et al. 2021 p.36–47). Thus reduced immunity, low organ function and co-existing co-morbidities are in combination responsible for increased risk of mortality in the advanced age groups.

Data analysis revealed that district wise mortality per million population was 790 and 735 in the district of Jammu and Srinagar respectively. On the national average, country has reported approximately 345 deaths per million (COVID-19 India Data Operations Group 2021) while globally this rate is 763. In contrast, USA reported a high rate of 2901 deaths per million population (Worldometer 2022). Comparative mortality figures for other states and UTs of India as of 31 January, 2022 reveal that 1,42,578 people died due to COVID-19 in Maharashtra; 25,827 deaths were reported in Delhi and 53,666 deaths in Kerala.

Co-morbidity has been reported to be as a major risk factor for COVID-19 mortality. In total, 79% of the COVID-19 deaths in Kashmir province had co-morbidity in comparison to 56% in Jammu province. The results of the current study are in congruity with those reported by Fang et al. and Wan et al. (Fang et al. 2020; Wan et al. 2020). Co-morbid conditions usually lead to reduced immune functions and most of these patients are taking multiple drugs concurrently which may also have adverse drug reactions. All these factors in conjunction with down regulation of immune functions increase the risk of mortality among these patients.

The results have further revealed that about one-third of the mortality in both the province of UT of Jammu & Kashmir was reported within 1–3 days since admission. About 10% of the patients with COVID-19 survived up to 15 days before finally succumbing to the disease. In Jammu province mortality within 24 hours since admission was 11.7% compared to 6.6% in Kashmir province. The time of death since admission points to a pattern wherein some patients who might be co-morbid and in higher age groups have died within hours/days of admission. Further the pattern shows that probably the people who were not co-morbid and younger in age battled COVID-19 for a longer time before losing the struggle. During intensive review of literature, the authors could not find any comparative study where time of death since admission as a variable was analysed.

Since only data of UT of Jammu & Kashmir for mortality was analysed, the results of the present study lack generalizability. Despite this, evidence from the current study is likely to advance epidemiologic information about role of various variables in COVID-19 mortality.

Conclusion

Male patients with COVID-19 were more at risk of mortality compared to females. 50–69 years of age group was the worst affected in the current study. About 80% of the deaths in Kashmir province were associated with co-morbidity compared to 56% of Jammu province. Time of death since admission was 1–3 days for about one-third of cases.

Recommendations

Vaccinated persons who are older, immunosuppressed or have other co-morbid conditions should receive targeted interventions including chronic disease management, precautions to reduce exposure, additional booster vaccine doses as advised and effective pharmacological interventions to mitigate risk for severe outcomes. Increasing vaccination coverage is a critical public health priority.

References

Ahmad S, Kumar P, Shekhar S, Saha R, Ranjan A, Pandey S. (2021) Epidemiological, clinical, and laboratory predictors of in-hospital mortality among COVID-19 patients admitted in a tertiary covid dedicated hospital, Northern India: A retrospective observational study. *Journal of Primary Care & Community Health*;12:21501327211041486. https://doi.org/10.1177/21501327211041486 PMID: 34427136

Andrews MA, Areekal B, Rajesh KR, Krishnan J, Suryakala R, Krishnan B et al. (2020) First confirmed case of COVID-19 infection in India: A case report. *Indian Journal of Medical Research*;151(5):490–92.

Bhopal SS, Bhopal R. (2020) Sex differential in COVID-19 mortality varies markedly by age. *Lancet*; 396(10250):532–3. https://doi.org/10.1016/S0140-6736(20)31748-7 PMID: 32798449

Biswas M, Rahaman S, Biswas TK, Haque Z, Ibrahim B. (2021) Association of sex, age, and comorbities with mortality in COVID-19 patients: A systematic review and meta-analysis. *Intervirology*;64:36–47.

Chauhan NK, Shadrach BJ, Garg MK, Bhatia P, Bhardwaj P, Gupta MK, et al. (2021). Predictors of clinical outcomes in adult COVID-19 patients admitted to a tertiary care hospital in India: An analytical cross-sectional study. *Acta Biomedicine*;92(3):e2021024. https://doi.org/10.23750/abm.v92i3.10630 PMID:34212921

Chen N, Zhou M, Dong X, Qu J, Gong F, Han Y, et al. (2020). Epidemiological and clinical characteristics of 99 cases of 2019 novel coronavirus pneumonia in Wuhan, China: A descriptive study. *Lancet;*395(10223):507–13.

Coronavirus statistics- Worldometer. https://www.worldometers.info/coronavirus/. Last accessed on 31st January, 2022.

COVID-19 India Data Operations Group, "Covid19India.org" (2021); https://www.covid19india.org/.

Djaharuddin I, Munawwarah S, Nurulita A, Ilyas M, Tabri NA, Lihawa N. (2021). Comorbidities and mortality in COVID-19 patients. *Gaceta Sanitaria*;35(S2):S530–S532.

ECDC. COVID-19 Situation Update Worldwide. As of July 13 2020. Situation updates worldwide 2020 [Internet]. Available from: https://www.ecdc.europa.eu/en/geographical-distribution-2019-ncov-cases.

ECDC. COVID-19 Situation Update Worldwide. As of November 17 2020' [Internet] https://www.ecdc.europa.eu/en/geographical-distribution-2019-ncov-cases

Fang L, Karakiulakis G, Roth M. (2020) Are patients with hypertension and diabetes mellitus at increased risk for COVID-19 infection? *Lancet Respiratory Medicine*;8(4):e21.

Galloway JB, Norton S, Barker RD, Brookes A, Carey I, Clarke BD, et al. (2020) A clinical risk score to identify patients with COVID-19 at high risk of critical care admission or death: An observational cohort study. *The Journal of infection*;81(2):282–8. https://doi.org/10.1016/j.jinf.2020.05.064 PMID: 32479771

Global Health 50/50. COVID-19: data disaggregated by age and sex. Date accessed: January 20, 2022. https://globalhealth5050.org/covid19/age-and-sex-data/

Huang C, Wang Y, Li X, Ren L, Zhao J, Hu Y et al. (2020) Clinical features of patients infected with 2019 novel coronavirus in Wuhan, China. *Lancet;*395(10223):497–506.

Ji Dong, Zhang Dawei, Xu Jing, et al. (2020). Prediction for progression risk in patients with COVID-19-19 pneumonia: The CALL score. *Clinical Infectious Diseases*;6(15):1393–1399. https://doi.org/10.1093/cid/ciaa414 PMID: 32271369

Joshi SR. (2020) Indian COVID-19 risk score, comorbidities and mortality. *The Journal of the Association of Physicians of India*;68(5):11–2. PMID: 32610858

Knight SR, Ho A, Pius R, Buchan I, Carson G, Drake TM, et al. (2020) Risk stratification of patients admitted to hospital with covid-19 using the ISARIC WHO clinical characterisation protocol: development and validation of the 4C Mortality Score. *BMJ* (Clinical research ed);370:m3339. https://doi.org/10.1136/bmj.m3339 PMID: 3290785500426 PMID: 32766541

Lavergne R, McGrail K. (2014) Amenable (or avoidable) mortality as an indicator of health system effectiveness. *Healthcare Policy = Politiques de Sante*. 2014;10:1 (January 1):8–9.

La Vignera S, Cannarella R, Condorelli RA, Torre F, Aversa A, Calogero AE. (2020). Sex-specific SARS-CoV2 mortality: Among hormonemodulated ace2 expression, risk of venous thromboembolism and hypovitaminosis D. *International Journal of Molecular Sciences*;21(8):2948. doi: 10.3390/ijms21082948.

Leng J, Goldstein DR. (2010) Impact of aging on viral infections. *Microbes and Infection*;12(14–15):1120–4.

Liang W, Liang H, Ou L, Chen B, Chen A, Li C, et al. (2020) Development and validation of a clinical risk score to predict the occurrence of critical illness in hospitalized patients with COVID-19. *JAMA Internal Medicine*;180(8):1081–9. https://doi.org/10.1001/jamainternmed.2020.2033 PMID: 32396163

Li X, Xu S, Yu M, Wang K, Tao Y, Zhou Y, et al. (2020). Risk factors for severity and mortality in adult COVID-19 inpatients in Wuhan. *The Journal of Allergy and Clinical Immunology*;146(1):110–8. https://doi.org/10.1016/j.jaci..04.006 PMID: 32294485

Nakakubo S, Suzuki M, Kamada K, Yamashita Y, Nakamura J, Horii H, et al. (2020) Proposal of COVID-19 clinical risk score for the management of suspected COVID-19 cases: A case control study. *BMC Infectious Diseases*;20(1):858. https://doi.org/10.1186/s12879-020-05604-4 PMID: 33208116

Peeri NC, Shrestha N, Rahman MS, Zaki R, Tan Z, Bibi S, Baghbanzadeh M, Aghamohammadi N, Zhang W, Haque U. (2020) The SARS, MERS and novel coronavirus (COVID-19) epidemics, the newest and biggest global health threats: What lessons have we learned? *International Journal of Epidemiology*;49:717–26.

Pradhan A, Olsson PE. (2020) Sex differences in severity and mortality from COVID-19: Are males more vulnerable? *Biology of Sex Differences*;11(1):53. https://doi.org/10.1186/s13293-020-00330-7 PMID: 32948238

Qian G-Q, Yang N-B, Ding F, Ma AHY, Wang Z-Y, Shen Y-F, et al. (2020). Epidemiologic and clinical characteristics of 91 hospitalized patients with COVID-19 in Zhejiang, China: A retrospective, multi-centre case series. *QJM*;113(7):474–81.

Rai D, Ranjan A, H A, Pandey S. (2021), Clinical and laboratory predictors of mortality in COVID-19 Infection: A retrospective observational study in a tertiary care hospital of Eastern India. *Cureus*;13(9):e17660. https://doi.org/10.7759/cureus.17660 PMID: 34646702

Salunke AA, Pathak SK, Dhanwate A, Warikoo V, Nandy K, Mendhe H, et al.(2020) A proposed ABCD scoring system for patient's self assessment and at emergency department with symptoms of COVID-19. *Diabetes & Metabolic Syndrome*;14(5):1495–501. https://doi.org/10.1016/j.dsx..07.053 PMID:32795741

Scully EP, Haverfield J, Ursin RL, Tannenbaum C, Klein SL. (2020) Considering how biological sex impacts immune responses and COVID-19 outcomes. *Nature Reviews Immunology*;20(7):442–7. https://doi.org/10.1038/s41577-020-0348-8 PMID: 32528136

Shang Y, Liu T, Wei Y, Li J, Shao L, Liu M, et al. (2020) Scoring systems for predicting mortality for severe patients with COVID-19. *EClinicalMedicine*;24:100426. https://doi.org/10.1016/j.eclinm.

Steinberg E, Balakrishna A, Habboushe J, Shawl A, Lee J. (2020) Calculated decisions: COVID-19 calculators during extreme resource-limited situations. *Emergency Medicine Practice*;22(4 Suppl):Cd1–cd5. PMID: 32259420

Varghese GM, John R, Manesh A, Karthik R, Abraham OC. (2020) Clinical management of COVID-19. *The Indian Journal of Medical Research*;151(5):401–10. https://doi.org/10.4103/ijmr.IJMR_957_20 PMID: 32611911

Wan Y, Shang J, Graham R, Baric RS, Li F. (2020) Receptor recognition by novel coronavirus from Wuhan: An analysis based on decadelong structural studies of SARS. *Journal of Virology*;94(7):e00127–20.

Williamson EJ, Walker AJ, Bhaskaran K, et al. (2020) Factors associated with COVID-19-related death using OpenSAFELY. *Nature*;584(7821):430–436. https://doi.org/10.1038/s41586-020-2521-4 PMID: 32640463

Xu X, Yu C, Qu J, Zhang L, Jiang S, Huang D et al. (2020) Imaging and clinical features of patients with 2019 novel coronavirus SARS-CoV-2. *European Journal of Nuclear Medicine and Molecular Imaging*;47(5):1275–80.

Yang X, Yu Y, Xu J, Shu H, Xia J, Liu H, et al. (2020). Clinical course and outcomes of critically ill patients with SARS-CoV-2 pneumonia in Wuhan, China: A single-centered, retrospective, observational study. *The Lancet Respiratory Medicine*;8(5):475–81.

Zhou F, Yu T, Du R, Fan G, Liu Y, Liu Z, et al. (2020) Clinical course and risk factors for mortality of adult inpatients with COVID-19 in Wuhan, China: A retrospective cohort study. *Lancet*;395(10229):1054–62. https://doi.org/10.1016/S0140-6736(20)30566-3 PMID: 32171076

Zhu N, Zhang D, Wang W, Li X, Yang B, Song J, et al. (2020) A novel coronavirus from patients with pneumonia in China. The New England Journal of Medicine;382:727–733.

22 Mental Health Assessment of Recovered COVID-19 Patients

A Case Study of Srinagar City

Aisha Dev, M. Imran Ganaie, Aabida,
Afshan Nabi and Ishtiaq A. Mayer

Introduction

Human history has witnessed many pestilences, pandemics, and plagues. From the plague of Athens in the 3rd century BC to the Spanish flu which killed more than 30 million people worldwide, mankind has always been vulnerable to stress and anxiety due to outbreaks of such events of pandemics and global diseases (Rosenwald, 2020). Thereby such pandemics are much beyond biological phenomena. They have psychosocial and economic implications that might long outlast the infection itself.

In December 2019, an outbreak of novel coronavirus pneumonia occurred in Wuhan (Hubei, China), and subsequently attracted worldwide attention by February 9, 2020 (Li et al., 2020, p.80–85; Rothan and Byrareddy, 2020). Ever since the outbreak of COVID-19 among the populace, an exponential rise in the public discourse around the information and concerns associated with the infection has been impacting global mental health (Chen et al., 2020; Dev et al., 2021). The unpredictability of the situation, the uncertainty of when to control the disease, its spread to different geographic locations, and the seriousness of the risk post-infection, together have aggravated the stressfulness of the situation (Onditi, 2020; Schomburg, 2021). These, along with publicly available risk analysis and misinformation through social media consumption and penetration where people are getting overloaded with unverified rumors regarding COVID-19, have heightened concern among the masses (Banerjee and Bhattacharya, 2020; Bao et al., 2020; Zandifar and Badrfam, 2020). Nevertheless, challenges and stress can act as triggers for the onset of or vulnerability of common mental health issues, such as anxiety and depression. Moreover, disasters disproportionately affect poor and vulnerable populations, and patients with serious mental illness may be among the hardest hit (Dar et al., 2017; Gourkhede et al., 2020).

The World Health Organisation (WHO) has also expressed its concern over the pandemic's mental health and psychosocial consequences. Psychologists and mental health professionals speculate that new measures such as self-isolation and quarantine have affected the usual activities, routines, and livelihoods of people may lead to an increase in loneliness, anxiety, depression, insomnia, harmful alcohol, and drug use, and self-harm or suicidal behavior (Kumar and Nayar, 2021,

DOI: 10.4324/9781003329459-22

p.1–2; Zhang et al., 2020; World Health Organization, 2020). In addition, people who are quarantined lose face-to-face connections and traditional social interventions, and this is a stressful phenomenon (Liu et al., 2020a; Moukaddam and Shah, 2020; Zandifar and Badrfan, 2020; Zhang et al., 2020, p.3–8).

The rapid spread of COVID-19 has impacted many spheres of human life and activities in India. The Indian society, by and large, is superstitious resulting from its low literacy rate, particularly in the rural populace which accounts for about 70% of the total population. During first and second phase of the pandemic, people would not claim the dead bodies of their home mates who had died due to COVID-19. The corpse was thrown into water bodies or on highways unclaimed by anyone. (Documentaries and news gallery of BBC, Aljazeera, and other National and global TV channels 2020–21) Considering the socio-cultural diversity, limited public health resources, increasing psychological co-morbidities, and a substantial number of vulnerable populations (e.g., homeless, migrants), the subcontinent is facing unprecedented challenges on all fronts (Banerjee and Bhattacharya, 2020). The drastic restrictions on economic activities have resulted in enormous economic losses and consequent loss of income and livelihood. Millions of people have lost their jobs. People employed in the informal and unorganized sectors are worst hit as they are struggling for, food, shelter, and livelihoods. This uncertainty consequently, led to serious mental health problems like depression, suicide, self-harm, etc. (Jungari, 2020; Kumar and Nayar, 2021, p.1–2).The lockdown may be an important strategy to break the chain of transmission however; it has also created boredom and monotony among office goers and children. In many households, children who end up staying indoors become restless and, in some cases, violent. Many households have even closed windows and doors and blocked all forms of ventilation in order to restrict the entry of virus into their apartments.

In Kashmir Himalaya, India, in particular, the long-standing political turmoil has had an unimaginable impact on the mental health of people (Varma, 2020). The ongoing pandemic has further deepened the problems, thereby burdening the mental health system of the state (Dalal et al., 2020; Varma et al., 2021). There has been an alarming rise in cases of depression, anxiety, and psychotic events in the Valley. According to a 2015 study by Médecins Sans Frontières and the Srinagar-based Institute of Mental Health and Neurosciences (IMHANS), nearly one in five people in Kashmir show symptoms of post-traumatic stress disorder. Separate research published by IMHANS and Action Aid in 2016 estimated that 11.3% of the population had a mental health disorder – higher than the national prevalence of around 7% (NusratSidiq, 2021). According to a study conducted by a United States-based organization, the National Center for Biotechnology Information (NCBI), the change in the status of Jammu and Kashmir from a state to a Union Territory after the abrogation of Article 370 in 2019, had an enduring psychological impact on the Kashmiri population (Shoaib and Arafat, 2020, p.65–66; TashafiNazir, 2021). Just six months after that, the coronavirus struck India and the whole country went under a strict lockdown. Doctors say that the situation has morphed into a severe psychological crisis. Besides, there has also been a surge in suicide cases in the Valley over the last six months, particularly during the second wave of COVID-19. The fear of unemployment, economic hardship, domestic violence, and indebtedness aggravated an already existing mental crisis amongst the people of the valley. According to the data available with the police, the Srinagar district alone recorded 29 suicide attempts since March in the year 2021. Twelve persons including six girls, three boys, and three elderly women died by suicide, jumping into river Jhelum in 2021. A majority of cases were reported in May, June, July, and August. This is the first time Srinagar has seen such a rise in suicide cases (Aabidwani, 2021). According to a report published by the National Human Rights Commission, 20,000 people have attempted suicide during the 14 years of socio-political turmoil in the Valley. About 3,000 of them have died and most of them were in the 16–25 age group. Official records from Shri Maharaja Hari Singh (SMHS) hospital, one of the major tertiary care hospitals in the city, reveal that over the past year the hospital has recorded over 500 suicide/attempted suicide cases from different parts of Kashmir with 172 males and 343 females. Data compiled by National Crime Records Bureau (NCRB) records around 6,000 cases of suicide in Kashmir between 1990 and 2019 (TashafiNazir, 2021). There is an immediate need for the development of mental health services in Kashmir accompanied by community participation, awareness programs, and mental health rehabilitation service. The present study attempts to highlight the mental health issues post-COVID-19 recovered patients of Srinagar City, Kashmir India.

Methodology

A total of 300 COVID-19 recovered patients is taken as a sample study. The sample set is divided into two groups (hospitalized and home quarantined). Out of 300 participants, ~20% who had severe symptoms who were hospitalized opted to participate, and ~80% with mild symptoms who were home quarantined opted to participate in the study. The hospitalized patients were primarily found to be from three hub Hospitals of Srinagar city dealing with the COVID-19- patients, i.e., (SKIMS, Chest Disease hospital, and JLNH, Rainawari). To evaluate post-COVID depression and anxiety, a questionnaire based on validated versions of the Patient Health Questionnaire-9 (PHQ-9) and the Generalized Anxiety Disorder-7 (GAD-7) was prepared online on Google forms. All the questions in the questionnaire had a set of four options that translated to an equivalent numeric score as per the PHQ-9 or GAD-7 scale. Given the strict COVID guidelines around social distancing and the personal safety of COVID survivors, online channels were preferred. Moreover, the method of snowball sampling was used to ensure extensive reach. Under this non-probability sampling technique, the target population was identified (hospitalized and home quarantined patients). The same was then used to identify future subjects from their acquaintances. The questionnaire was distributed through an online method to the target participants using social networks, mostly WhatsApp and emails. Through these first respondents, the reach of the questionnaire was further extended with the participants sharing the same with their acquaintances. Telephonic interviews were also recorded. After the responses were received, a cumulative score was calculated for each respondent for each scale.

The questionnaire also covered the demographic aspects of the respondents. The Chi-square test is used to compare the prevalence of mild to severe mental health problems between hospitalized and home quarantined patients. To analyze the severity of mental health problems amongst the COVID-19 survivors with depression and anxiety, logistic regression is employed to identify statistically significant variables from amongst several variables like (age, gender, education, income level, pre-existing health problems, re-testing positive for COVID-19, recovery period).

Results

For all the COVID-19 survivors (median age: 40 years, inter quartile range, IQR (Inter-quartile range): 29–50 years; 65.75% female and 33.15% males), the median duration of the follow-up period was 9–12 days. During this period, 37.02% were found to have clinically defined depression (i.e., a score of at least ten on PHQ, as shown in Table 22.1, and 24.31% to have clinically defined anxiety (i.e., a score of at least ten on GAD-7, as shown in Table 22.2).

Comparison between the 79.67% home quarantined patients and 20.33% hospitalized patients (Table 22.3) revealed that the risk of severe depression (relative risk, RR =1.308, 95% CI: 0.77 – 2.2, as shown in Table 22.4), as well as severe anxiety levels relative risk, RR = 3.35, 95% CI: 2.4 – 4.6, as shown in Table 22.5), was higher in home quarantined patients.

Among the 37.02% of survivors with depression, the risk of severe conditions (i.e., score of at least 10 on the PHQ) was more pronounced in respondents who retested positive for COVID-19 (OR: 20.8, $p < 0.03$ at 95% confidence level) and in respondents with any pre-existing health condition (diabetes, hypertension, heart problems) (OR: 1.1, 95% CI, $p < 0.003$). Amongst these respondents, it was also observed that resilience towards aggravated depression severities was common in

Table 22.1 PHQ-9 score distribution of respondents

PHQ-9 (< 5) minimal depression	PHQ-9 (5–9) mild depression	PHQ-9 (10–14) moderate depression	PHQ-9 (15–19) moderately severe depression	PHQ-9 (20+) severe depression
39.23%	23.76%	17.68%	15.47%	3.87%

Source: Primary data from PHQ-9 questionnaire.

Table 22.2 GAD-7 score distribution of respondents

GAD-7 (<5) normal	GAD-7 (5-9) mild anxiety	GAD-7 (10-14) moderate anxiety	GAD-7 (14+) severe anxiety
54.14%%	21.55%	22.10%	2.21%

Source: Primary data from GAD-7 questionnaire

Table 22.3 Classification of respondents into hospitalized and home quarantined

Total Respondents	Hospitalized	%Hospitalized	Home Quarantined	% Home Quarantined
300	61	20.33%	239	79.67%

Source: Primary data collected from survey.

Table 22.4 Relative risk estimate for depression

	Value	95% Confidence Interval	
		Lower	Upper
Odds ratio for home quarantined (no/yes)	0.665	0.305	1.451
For cohort severe PHQ = no	0.870	0.679	1.113
For cohort severe PHQ = yes	1.308	0.765	2.236
N of valid cases	300		

Source: Primary data analyzed using Chi-square test.

Table 22.5 Relative risk estimate for anxiety

	Value	95% Confidence Interval	
		Lower	Upper
Odds ratio for home quarantined (no/yes)	0.065	0.025	0.167
For cohort severe GAD = no	0.216	0.103	0.452
For cohort severe GAD = yes	3.351	2.442	4.599
N of valid cases	300		

Source: Primary data analyzed using chi-square test.

Males (Gender – 1) (OR: 0.2, 95% CI, $p < 0.001$) and those who had been hospitalized for COVID care (OR: 0.09, 95% CI, $p < 0.003$). Details of these relationships between various risk factors with depression are shown in Table 22.6.

Among the 24.31% of survivors with anxiety, the risk for severe conditions (i.e., a score of at least ten on GAD-7 was more prevalent in respondents who took longer to recover from COVID-19 (>12 days) (OR: 18.9, 95% CI, $p < 0.006$) and in those who had any pre-existing health condition (diabetes, hypertension, heart problems) (OR: 4.6, 95% CI, $p < 0.02$). Details of these relationships between various risk factors with anxiety are shown in Table 22.7.

Of the analyzed parameters, across anxiety and depression scales, it was found that age, occupation, education level and income levels did not play a significant role in impacting the overall mental health of the recovered COVID patients. These did not exhibit a statistically significant relationship with either of the factors under study – depression, anxiety.

Table 22.6 Relationship of risk factors with PHQ-9 (depression) score

PHQ Scale	Std. Error	Significance	Odds Ratio
Gender	**1.165**	**0.001**	**0.202**
Age	0.44	0.384	0.962
Education	0.384	0.788	1.109
Disposable income	0.356	0.560	0.813
Pre-existing condition	**0.371**	**0.003**	**1.112**
Retested positive	**1.389**	**0.029**	**20.876**
Recovery days	0.557	0.279	1.827
Hospitalized	**1.578**	**0.003**	**0.09**

Source: Primary data analyzed using logistic regression.

Table 22.7 Relationship of risk factors with GAD-7 (anxiety) score

GAD scale	Std. error	Significance	Odds RATIO
Gender	1.533	0.201	0.137
Age	0.059	0.929	1.005
Education	0.780	0.300	0.446
Disposable income	0.601	0.974	0.981
Pre-existing condition	**0.371**	**0.019**	**4.587**
Retested positive	2.633	0.141	0.021
Recovery days	**1.067**	**0.006**	**18.903**
Hospitalized	1.957	0.199	0.081

Source: Primary data analyzed with logistic regression.

Discussions

The COVID-19 pandemic has caused severe worldwide mental health problems. Delivery of mental health services has been adversely affected in all countries, but the effects have been compounded in the Kashmir valley due to the enduring political turmoil (Jong et al., 2006). Kashmir's seven-month communications blackout, which was enforced by the central government of India due to a change in the constitution, was lifted just before the COVID-19 pandemic began, further adding to the trauma to Kashmir valley (Shoaib and Arafat, 2020, p.65–66). Out of the 300 participants, both hospitalized and home quarantined, 37.02% were found to have clinically defined depression (PHQ-9 score > 10) and 24.31% were found to be having clinical defined anxiety (GAD-7 score > 10).

Factors like Retesting positive for COVID-19 and pre-existing medical health conditions (diabetes, hypertension, heart problems) are found to be a statistically significant factor responsible for severe depression amongst the respondents (OR: 20.8, $p < 0.03$ at 95% confidence level and OR: 1.1, 95% CI, $p < 0.003$ respectively). For severe anxiety, longer time to recover (OR: 18.9, 95% CI, $p < 0.006$) and prevalence of pre-existing medical conditions (OR: 4.6, 95% CI, $p < 0.02$)were found to be significant. Pre-existing medical conditions emerged as one common factor in patients suffering from both depression and anxiety. Patients who even reported mild symptoms but had some pre-existing medical conditions had the added burden of taking greater precautions. Moreover, the information discourse around the severity of COVID in patients with medical conditions may have only aggravated their anxiousness, which is reflected in them scoring higher on GAD-7 as well as PHQ-9. This is further validated by a similar study carried out on 1,099 COVID-19 patients which stated that patients with at least one comorbidity (Hypertension, diabetes, and coronary heart diseases being most common) are associated with poorer clinical outcomes (Mao et al., 2020). Similarly, retesting positive for COVID-19 aggravated the stress and anxiety levels of the patients thereby affecting the mental health of the people. This called out for extended isolation and loneliness from their family and loved ones and a longer recovery period. Studies have revealed that the fear of loneliness and isolation has resulted in deep mental stress and depression amongst the COVID-19 patients.

Results also yielded that in comparison to females, males showed resilience towards aggravated depression severities (OR: 0.2, 95% CI, $p < 0.001$). A recent study, carried out by four doctors from Govt Medical College Srinagar from the department of community medicine revealed that 16% of adolescents were found to be clinically depressed with 14.5% boys and 18% girls. Sleep deprivation being a common symptom. Besides depression, anxiety was found in 40% of adolescents with 14% of boys and 27.5% of girls. Reasons being a past history of COVID-19 and hospitalization of family members due to COVID-19 (Zain Bin Shabir, 2022, p.1 and 6). These findings largely correspond with the findings of present research paper. In the wake of the pandemic, almost every household turned into an office, school, all in one. Multiple homes saw an increase in workload during the lockdown. Social and economic factors can put women at greater risk of poor mental health than men. Besides mood-related issues, emotional outbursts in women especially panic, fear, avoidance, and fear of meeting other people, fear of death, getting isolated, and fear of not getting essential household and other items may have their psychological manifestations. Also, it is seen that strict lockdowns have led to an increase in cases of domestic violence where women and children who live with domestic violence have no escape from their abusers during quarantine (Kumar and Nayar, 2021, p.1–2). Similar studies were carried out in Wuhan china, the UK by (Li and Wang, 2020, p.80–85; Liu et al., 2020) in order to investigate the prevalence of psychiatric disorders during the COVID-19 pandemic peak, it was found that post-traumatic stress symptoms were prevalent more in women.

Likewise, higher resilience was also observed amongst those who had been hospitalized for COVID care (OR: 0.09, 95% CI, $p < 0.003$) in comparison to those home quarantined. On interviewing the recovered patients who were hospitalized, it was stated that the feeling of being taken care of by the medical professional and other staff creates a sense of relief in them. Being away from family puts negative stress on them but at the same time, their safety was their utmost priority. This is also seen in the results as the relative risk of severe depression as well as anxiety is also higher in home quarantined patients v/s the hospitalized (relative risk, RR =1.308, 95% CI: 0.77 – 2.2; and relative risk, RR = 3.35, 95% CI: 2.4 – 4.6 respectively.

Conclusion

The study documents that the mental health problems of the COVID-19 recovered patients are more significant in the home quarantined patients than the hospitalized, as the latter proved to be more resilient. Four risk factors (Retesting positive for COVID-19, pre-existing medical condition (diabetes, hypertension, heart problems), longer time of recovery (> 12 days), and gender) played a significant role in severe mental health problems. The psychological behavior of the population during the outbreak of any infectious disease plays a critical role in shaping both the spread of the disease and the occurrence of emotional distress and social disorder during and after the outbreak. Therefore, the need for mental health support continues to be of utmost importance. The study suggests that immediate intervention can and should be done to minimize the psychological and psychiatric effects of the COVID-19 pandemic. An urgent need for researchers, clinicians, and policymakers for devising policies coupled with a well-equipped telepsychiatry service system is the need of the hour to deal with the burning problem of deteriorating mental health in Kashmir valley. Nevertheless, mental illness should not be treated as a taboo, especially in societies with conflicts. Destigmatizing mental health conversations in households would play a crucial role in helping people struggling with mental health issues to reach out to more people and seek help. This approach will boost the accessibility and affordability of mental health interventions with timely diagnosis and improve the follow-up for treatment.

The need for mental health support continues to be of utmost importance. An urgent need for researchers, clinicians, and policymakers for devising policies coupled with a well-equipped telepsychiatry service system is the need of the hour to deal with the burning problem of deteriorating mental health of the Kashmir valley.

References

Banerjee, D., & Bhattacharya, P. (2020). "Pandemonium of the pandemic": Impact of COVID-19 in India, focus on mental health. *Psychological Trauma: Theory, Research, Practice, and Policy, 12*(6), 588.

Bao, Y., Sun, Y., Meng, S., Shi, J., & Lu, L. (2020). 2019-nCoV epidemic: address mental health care to empower society. *The Lancet, 395*(10224), e37–e38.

Chen, Q., Liang, M., Li, Y., Guo, J., Fei, D., Wang, L., ... & Zhang, Z. (2020). Mental health care for medical staff in China during the COVID-19 outbreak. *The Lancet Psychiatry, 7*(4), e15–e16.

Dalal, P. K., Roy, D., Choudhary, P., Kar, S. K., & Tripathi, A. (2020). Emerging mental health issues during the COVID-19 pandemic: An Indian perspective. *Indian Journal of Psychiatry, 62*(Suppl 3), S354.

Dar, K. A., Iqbal, N., & Mushtaq, A. (2017). Intolerance of uncertainty, depression, and anxiety: Examining the indirect and moderating effects of worry. *Asian Journal of Psychiatry, 29*, 129–133.

de Jong, K., vdKam, S., Fromm, S., van Galen, R., & Kemmere, T. (2006). Kashmir: Violence and health. MédecinsSansFrontieres, Amsterdam, The Netherlands

Dev, A., Ganaie, M. I., Mayer, M. A., & Singh, H. (2021). Mental health response of people towards COVID-19 outbreak in Kashmir valley. *Wutan Huatan Jisuan Jishu, 17*(1), pp 151–161, ISSN: 1001-1749.

Gourkhede, D. P., Ravichandran, K., Kandhan, S., Ram, V. P., Dhayananth, B., Megha, G. K., & Kumar, M. S. (2020). COVID-19: Mental health issues and impact on different professions. *International Journal of Current Microbiology and Applied Sciences*, *9*(7), 2994–3013.

Kumar, A., & Nayar, K. R. (2021). COVID 19 and its mental health consequences. *Journal of Mental Health*, *30*(1), 1–2.

Jungari, S. (2020). Maternal mental health in India during COVID-19. *Public Health*, *185*, 97.

Mao, R., Liang, J., Shen, J., Ghosh, S., Zhu, L. R., Yang, H., ... & Chen, M. H. (2020). Implications of COVID-19 for patients with pre-existing digestive diseases. *The Lancet Gastroenterology & Hepatology*, *5*(5), 425–427.

Moukaddam, N., & Shah, A. (2020). Psychiatrists beware! The impact of COVID-19 and pandemics on mental health. *Psychiatric Times*, *37*(3).

Li, L. Z., & Wang, S. (2020). Prevalence and predictors of general psychiatric disorders and loneliness during COVID-19 in the United Kingdom. *Psychiatry research*, *291*, 113267.

Li, J. Y., You, Z., Wang, Q., Zhou, Z. J., Qiu, Y., Luo, R., &Ge, X. Y. (2020). The epidemic of 2019-novel-coronavirus (2019-nCoV) pneumonia and insights for emerging infectious diseases in the future. *Microbes and Infection*, *22*(2), 80–85.

Liu, C. H., Zhang, E., Wong, G. T. F., & Hyun, S. (2020a). Factors associated with depression, anxiety, and PTSD symptomatology during the COVID-19 pandemic: Clinical implications for US young adult mental health. *Psychiatry Research*, *290*, 113172.

Liu, N., Zhang, F., Wei, C., Jia, Y., Shang, Z., Sun, L., ... & Liu, W. (2020b). Prevalence and predictors of PTSS during COVID-19 outbreak in China hardest-hit areas: Gender differences matter. *Psychiatry Research*, *287*, 112921.

Liu, S., Yang, L., Zhang, C., Xiang, Y. T., Liu, Z., Hu, S., & Zhang, B. (2020c). Online mental health services in China during the COVID-19 outbreak. *The Lancet Psychiatry*, *7*(4), e17–e18.

Onditi, F., Obimbo, M. M., Kinyanjui, S. M., & Nyadera, I. N. (2020). Rejection of containment policy in the management of COVID-19 in Kenyan slums: Is social geometry an option?

Rosenwald, M. S. (2020). History's deadliest pandemics, from ancient Rome to modern America. *Washington Post*, 7.

Rothan, H. A., & Byrareddy, S. N. (2020). The epidemiology and pathogenesis of coronavirus disease (COVID-19) outbreak. *Journal of Autoimmunity*, *109*, 102433.

Schomburg, L. (2021). Selenium deficiency due to diet, pregnancy, severe illness, or COVID-19—A preventable trigger for autoimmune disease. *International Journal of Molecular Sciences*, *22*(16), 8532.

Shoaib, S., & Arafat, S. Y. (2020). Mental health in Kashmir: Conflict to COVID-19. *Public Health*, *187*, 65–66.

Thibaut, F., & van Wijngaarden-Cremers, P. J. (2020). Women's mental health in the time of Covid-19 pandemic. *Frontiers in Global Women's Health*, *1*, 17.

Varma, S. (2020). *The occupied clinic: Militarism and care in Kashmir*. Duke University Press, Durham, NC.

Varma, R., Das, S., & Singh, T. (2021). Cyberchondria amidst COVID-19 pandemic: challenges and management strategies. *Frontiers in Psychiatry*, *12*, 399.

World Health Organization. (2020). Mental health and psychosocial considerations during the COVID-19 outbreak, 18 March 2020 (No. WHO/2019-nCoV/MentalHealth/2020.1). World Health Organization.

Zandifar, A., &Badrfam, R. (2020). Iranian mental health during the COVID-19 epidemic. *Asian Journal of Psychiatry*, June 51. DOI: 10.1016/j.ajp.2020.101990.

Zhang, J., Wu, W., Zhao, X., & Zhang, W. (2020). Recommended psychological crisis intervention response to the 2019 novel coronavirus pneumonia outbreak in China: A model of West China Hospital. *Precision Clinical Medicine*, *3*(1), 3–8.

Web Sources

Nazir, T. (2021). Covid, lockdowns aggravated the already existing mental health crisis in Kashmir. *The Logical Indian.* https://thelogicalindian.com/mentalhealth/covid-19-pandemic-lockdowns-aggravated-already-existing-mental-health-crisis-in-kashmir-29997

Shabir, BZ. (2022, April 14). 16% of teens battling coronavirus triggered depression. *Kashmir observer.* https://epaper.kashmirobserver.net/epaper/edition/1289/kashmir-observer-april/page/1.

Sidiq, N. (2019, December 23). Kashmir's mental health crisis goes untreated as clampdown continues. *The New Humanitarian.* https://www.thenewhumanitarian.org/news/2019/12/23/Kashmir-conflict-mental-health

Wani, AB. (2021, September 10). What is behind the alarming rise in suicides in Srinagar? *Citizen Matters.* https://citizenmatters.in/conflict-and-covid-causing-rise-in-suicides-in-srinagar-27090

23 Conclusion and Suggestions

Rais Akhtar

A comparative study between historical and contemporary scenario of environmental perspective of health and disease is an enthralling area of research which highlights the pattern of environmental conditions, people's adjustment with the environment and people's mobility, and the occurrence of health problems as a result of man's interaction with the environment. *Airs, Waters and Places,* by Hippocrates published in 5th century BC laid more thrust on environmental factors which play an important role in regional variations in the distribution of diseases. This is relevant with the outbreaks of cholera in Kashmir during 19th century and the recent occurrences of cardiovascular and COVID-19 in Jammu & Kashmir and Ladakh. Jammu & Kashmir and Ladakh are geographically, both physical and human geographies, is an interested region for study.

During the past 160 years, in Kashmir and Ladakh, contributing factors which have shaped the modern public health system include missionary doctors as well as spread of education which resulted in people began accepting modern allopathic medicine. With the growth of scientific medical knowledge about sources and means of controlling communicable, waterborne and region-based (for example Kangri cancer) diseases resulted in the growth of public awareness and acceptance of disease control measures, particularly cholera and smallpox in Kashmir and in Ladakh, although writing by Henry Caley, Chapter 3 suggests that there was no case of smallpox in the year 1867 but that "ten years ago it spread through the whole country, and killed numbers". The most common ailments prevalent currently in the Ladakh division are arthritis followed by peptic disorders, whereas diseases like diabetes, TB, carcinomas, heart ailments are found with low intensity.

Example of disease outbreaks in the region includes and cholera and smallpox in the past and COVID-19 outbreaks and its diffusion, cancer and heart ailments in contemporary scenario are important examples in the context of Jammu & Kashmir and Ladakh.

Such regional geographies are being studied in various developing countries particularly mountainous regions. In the light of this, I collected studies encompassing 19th century and on contemporary scenarios on various aspects of environment and health, and edited them in order to publish it in a book form. It took a little more than two years to complete.

DOI: 10.4324/9781003329459-23

The edited book *Historical and Contemporary Scenarios of Environment and Health in Jammu & Kashmir and Ladakh* comprises chapters encompassing cholera outbreaks in Kashmir, practice of *Kangri* and the Kangri burn cancer, contributions of medical missionaries in health services provision in Kashmir, changing disease ecology of Leh, traditional medical therapy in Ladakh, environmental compulsions and pesticide application in Kashmir and its impacts on orchardists, urbanization and quality of life in Srinagar, vector borne diseases, extreme weather events (Flooding in Kashmir in 2013), people's perception of urban health hazards, spatial pattern of diseases, health care behaviour, food insecurity and COVID-19 spatial pattern in Jammu & Kashmir and Ladakh.

It has rightly been said that environment and health are intrinsically linked. Where we live, and work directly influences our health perceptions, and the health services we can access. The social and natural environments influence our health and wellbeing, and are directly relevant to the formulation of regional health care policy. Spatial location of places and the correspondence between places results in shaping environmental/health risks. For example, locating health care facilities, targeting public health strategies or monitoring disease outbreaks, and its diffusion, all have a geographic context.

Thus there is need to carry out similar case studies in different regions in developing countries of Asia, Africa and Latin America, in order to understand environment and human health scenario, to help develop holistic regional health care planning.

Index

Note: **Bold** page numbers refer to tables; *italic* page numbers refer to figures and page numbers followed by "n" denote endnotes.

Taylor & Francis Group
an **informa** business

Taylor & Francis eBooks

www.taylorfrancis.com

A single destination for eBooks from Taylor & Francis
with increased functionality and an improved user
experience to meet the needs of our customers.

90,000+ eBooks of award-winning academic content in
Humanities, Social Science, Science, Technology, Engineering,
and Medical written by a global network of editors and authors.

TAYLOR & FRANCIS EBOOKS OFFERS:

A streamlined
experience for
our library
customers

A single point
of discovery
for all of our
eBook content

Improved
search and
discovery of
content at both
book and
chapter level

REQUEST A FREE TRIAL
support@taylorfrancis.com

Routledge
Taylor & Francis Group

CRC Press
Taylor & Francis Group

For Product Safety Concerns and Information please contact our EU
representative GPSR@taylorandfrancis.com
Taylor & Francis Verlag GmbH, Kaufingerstraße 24, 80331 München, Germany

www.ingramcontent.com/pod-product-compliance
Lightning Source LLC
Chambersburg PA
CBHW060240220326
41598CB00027B/3996

* 9 7 8 1 0 3 2 3 5 9 4 7 2 *